Heike Schröder

Negotiating the Kyoto Protocol

Münchener Beiträge zur Geschichte und Gegenwart der internationalen Politik

herausgegeben von

Peter J. Opitz

unter Mitwirkung von

Mir A. Ferdowsi

und

Dietmar Herz

BAND 9

LIT

Heike Schröder

Negotiating the Kyoto Protocol

An analysis of negotiation dynamics in international negotiations

LIT

Umschlagbild: Fukui Asahido Co., Ltd.

for my parents

Die Deutsche Bibliothek – CIP-Einheitsaufnahme

Schröder, Heike:
Negotiating the Kyoto Protocol : An analysis of negotiation dynamics in international negotiations / Heike Schröder. – Münster : LIT, 2001
(Münchener Beiträge zur Geschichte und Gegenwart der internationalen Politik ; 9.)
ISBN 3-8258-5446-9

Grevener Str. 179 48159 Münster Tel. 0251-23 50 91 Fax 0251-23 19 72

Distributed in North America by:

Transaction Publishers
New Brunswick (U.S.A.) and London (U.K.)

Transaction Publishers
Rutgers University
35 Berrue Circle
Piscataway, NJ 08854

Tel.: (732) 445 - 2280
Fax: (732) 445 - 3138
for orders (U. S. only):
toll free (888) 999 - 6778

For my parents

"And one should bear in mind that there is nothing more difficult to execute, nor more dubious of success, nor more dangerous to handle than to introduce a new order of things: for a reformer has all those who profit from the old system as his enemies, and only lukewarm allies in all those who might profit from the new system. These allies are lukewarm partly from fear of their adversaries, who have the laws in their favour; and partly from the incredulity of mankind, who do not truly believe in anything new until they have had actual experience of it."

Niccolò Machiavelli, The Prince[1]

---

1 Unpublished translation by Campion Carruthers, September 1999.

Table of Contents

TABLES AND FIGURES

Foreword

The world is becoming a more dangerous place. The great challenges we face in the new millennium are not the aversion of wars between nation-states, but rather new types of conflict, such as mass proliferation of atomic, biological and chemical weaponry, international terrorism, drug trafficking, fears of the adverse consequences of globalisation, as well as the rise of many new global environmental problems. They are pressing global problems that require innovative global solutions.

To date, such transborder problems have been approached by establishing intergovernmental processes to develop procedures for international action, aimed to solve the problem at hand. However, this action has been greatly handicapped as the negotiated procedures have reflected little more than tainted national interests, and it has thus remained slow and cumbersome.

It is therefore necessary to develop and improve international policy structures that ultimately achieve a balancing act between safeguarding the sovereignty of states in choosing their implementation measures according to national circumstances, on the one hand, and establishing an effective compliance regime in order to assure the desired results, on the other.

Climate change is an area where a unique treaty has been negotiated – the Kyoto Protocol to the United Nations Framework Convention on Climate Change, adopted in Kyoto, Japan, in December 1997. It contains legally-binding emission reduction targets and various flexibility measures to help to efficiently implement its commitments.

The climate change issue has posed unprecedented challenges to policy makers. It is a global problem of unique complexity because it involves a wide spectrum of constituencies, including sovereign states, international organisations as well as business, environmental and local interest groups, all pursuing highly conflicting interests. The Kyoto Protocol therefore represents a milestone in the evolution of multilateral diplomacy as a means of tackling pressing global problems.

Academic literature has hitherto neglected the analysis of negotiation dynamics during international conferences. It is the responsibility of political scientists to investigate the determining factors that shape negotiation processes.

These include not only the positions, interests and strategies of crucial actors but also circumstances such as the general framework of the conference, public expectations, delegation structures and, not least, the power and personality of the chair of the respective negotiating body. Whilst there has been considerable focus on the outcomes of negotiation processes, the examination of negotiation dynamics has not yet attracted much scholarly attention.

This volume represents a case study that will address this lack of attention by offering an analysis of the negotiations of the Kyoto Protocol. In doing so, it will investigate the key factors instrumental to this decision-making process by examining the positions, interests and strategies of the three most crucial actors in the process, i.e., the European Union, the United States of America and Japan. Furthermore, it will explore the additional dynamics of the conference that shaped the final outcome. Its methodology can serve as a comprehensive model for future studies of the dynamics of international negotiation processes.

The author has first-hand insights into the climate change proceedings because she has had the invaluable opportunity to witness the COP3 negotiations at Kyoto, and subsequent meetings, as a note-taker for the UNFCCC Secretariat. She is thus well-acquainted with the proceedings and style of work of the UNFCCC process. In addition, the author has profited from her experience in a Model United Nations simulation, enabling her to use her newly developed understanding of the subtle dynamics of international bargaining to her analysis of the Kyoto negotiation process. Furthermore, she has gained invaluable insights into the international environmental policy debate by working as a Research Assistant at the Human Dimensions Programme for Global Environmental Change (IHDP) for over two years. These experiences make her a competent judge and commentator of the Kyoto deliberations.

Dietmar Herz
University of Erfurt, Autumn 2000

Acknowledgements

I am very grateful to Prof. Dr. Dr. h.c. Karl Kaiser and Prof. Dr. Dietmar Herz for supervising this study.

I thank Dr. Larry Kohler, Dr. Juha I. Uitto, Glen Paoletto and Vitaly Matsarski for invaluable insights into environmental and, above all, climate-related issues, and for enabling me to witness the final round of negotiations at the Kyoto Conference in December 1997 as a note-taker for the UNFCCC Secretariat. I thank Kushilani Wijesiri, UNFCCC Library, for her untiring help and assistance in gathering relevant documents and publications.

I am indebted to Joanna Depledge for reviewing the text and correcting many substantive errors. Brendan Barrett and Hee Boon Foo have enriched the manuscript in substance, style and language. Eva Maczuga and Stefanie Teggemann have made many improvements throughout the writing process, and I thank them for all their excellent ideas and encouragement. Finally I thank Martin Hale for greatly improving the language of the manuscript.

1. INTRODUCTION

Objectives

The 1997 Kyoto Conference[2] was the culmination of a two-and-a-half-year negotiation process. Its outcome – the Kyoto Protocol – represents one of the most ambitious and controversial global environmental undertakings to date[3]. If effectively implemented, it will represent the first step towards complete restructuring of the world's energy sector, despite the lack of *full* scientific evidence that human-induced climate change is taking effect.

Adopted through consensus by 159 states, each representing widely divergent and highly conflicting interests and agendas, the Kyoto Protocol is a remarkable diplomatic achievement. It was forged from conflicting national interests amidst a background of high public expectations. This public attention was focused primarily on the European Union, the United States and Japan, the main polluting nations, to strike the final deal. Every day at Kyoto was highlighted by new dynamics and compromises, and the outcome of the negotiations remained uncertain until the final hours of the conference.

This study will investigate the respective positions, strategies and underlying interests of the above-mentioned three key players during this process, and analyse the negotiation dynamics of the Kyoto Conference. It will endeavour to explore the relationship between national interests and other factors, such as the

---

2 The Kyoto Conference refers to the third session of the Conference of the Parties to the UN Framework Convention on Climate Change (COP3), held from 1-11 December 1997, in Kyoto, Japan.

3 While some analysts, such as Benedick and Cooper, have criticised the Kyoto Protocol for constituting a wrong approach to the problem, others have praised the Protocol as a unique achievement in the realms of international environmental law. Richard E. Benedick (1998), 'How to Salvage the Kyoto Protocol', Weathervane, March 1998, at: <http://www.weathervane.rff.org/pop/pop5/benedick.html>. Richard N. Cooper (1998), 'Towards a Real Global Warming Treaty', pp. 66-79, in: *Foreign Affairs*, Vol. 77, No. 2, March/April 1998.

personality of the chair, the mounting fatigue of delegates as well as high public expectations, all of which had an impact on the dynamics of the Conference.

The political process up to Kyoto is well documented[4] and some of the crucial issues included in the Protocol, such as the flexibility mechanisms, have already been extensively covered considering the short time of their conceptual existence.[5] However, until now, the literature on the negotiations at COP3 remains limited.[6]

4 See, for example, Paul Brown (1996), *Global Warming, Can Civilization Survive?*, London; Mathew Paterson (1996), *Global Warming and Global Politics*, London; Jill Jäger and Tim O'Riordan (1996), 'The History of Climate Change Science and Politics', pp. 1-31, in: Jill Jäger and Tim O'Riordan (eds.), *Politics of Climate Change*, London; Reinhard Loske (1996), *Klimapolitik im Spannungsfeld von Kurzzeitinteressen und Langzeiterfordernissen*, Marburg; and Hermann Ott (1997) 'Das internationale Regime zum Schutz des Klimas', pp. 201-218, in: Thomas Gehring and Sebastian Oberthür (eds.), *Internationale Umweltregime, Umweltschutz durch Verhandlungen und Verträge*, Opladen.

5 See, for example, Thomas C. Heller (1999), 'Environmental Realpolitik: Joint Implementation and Climate Change', April 1999, at: <http://www.law.indiana.edu/glsj/vol3/no2/heller.html>; Fanny Missfeldt (1998), 'Flexibility Mechanisms: Which Path to Take after Kyoto?' pp. 128-139, in: *Review of European Community & International Environmental Law*, Volume 7, Issue 2, 1998; Michael Toman and Marina Cazorla (1998) 'The Clean Development Mechanism: A Primer', Resources for the Future, September 1998, at: <http://teriin.org/treat/docs/res2.htm>; Jacob Werksman (1998) 'The Clean Development Mechanism: Unwrapping the 'Kyoto Surprise'', pp. 147-158, in: *Review of European Community & International Environmental Law*, Volume 7, Issue 2, 1998; Michael Grubb (1998), 'International Emissions Trading under the Kyoto Protocol: Core Issues in Implementation', pp. 140-146, in: *Review of European Community & International Environmental Law*, Volume 7, Issue 2, 1998; and P. Cullet and K. Kameri-Mbote (1998), 'Joint Implementation and Forestry Projects: Conceptual and Operational Fallacies', pp. 393-408, in: *International Affairs*, Vol. 74, No. 2, 1998.

6 Publications that cover at least some aspects of the COP3 negotiation process are Michael Grubb *et al.* (1999), *The Kyoto Protocol, A Guide and Assessment*, London; Hermann Ott (1998), 'The Kyoto Protocol to the UN Framework Convention on Climate Change – Finished and Unfinished Business', Wuppertal Institut für Klima, Umwelt, Energie GmbH, February 1998, at: <http://www.wupperinst.org/Publikationen/Kyoto_Protokoll.html>; Farhana Yamin (1998), 'The Kyoto Protocol: Origins, Assessment and Future Challenges', pp. 113-127, in: *Review of European Community & International Environmental Law*, Volume 7, Issue 2, 1998; Brendan F. D. Barrett and W. Bradnee Chambers (1998 Consultation

Structure

This study will examine the key aspects that ultimately shaped the outcome of the Kyoto Protocol. Moving from examination of the foundations of the climate change issues to investigating the key players and their political motivations, it will then analyse the national strategies and negotiation dynamics that, in the end, shaped the outcome of the conference.

CHAPTER ONE outlines the development of climate change research and the formation of an international scientific community related to the topic. It will then examine the politicisation of the climate change issue and conclude by exploring the UN Framework Convention on Climate Change, which constituted the first step in the political process of mitigating climate change and its adverse effects.

CHAPTER TWO develops the role of the three main players, the European Union, the United States and Japan, in the climate change process. Because of the unique structure of the EU, namely that it is not an individual state, special reference will be given to its particular nature as an actor in the process. The first section, 'Macro-Perspective', evaluates the climate change policies of the EU, US and Japan, and substantiates their negotiating positions for COP3. The second section, 'Micro-Perspective', compares and contrasts the various interests and structures that have shaped negotiations at national and international levels, including respective industrial interests, governmental frameworks, and energy supply structures between the EU, US and Japan.

CHAPTER THREE analyses the Kyoto negotiations up until and including the Kyoto Conference in December 1997. It first outlines the pre-Kyoto negotiation process that started with the Berlin Mandate in 1995. It will then explore the individual issues of the Kyoto Protocol and investigate how each of these issues was negotiated to determine the final outcome of the Conference. Finally it analyses the outcome of the negotiations, with special reference to the underlying strategies of the major players and the dynamics of the Conference.

Draft), 'Primer on Scientific Knowledge and Politics in the Evolving Global Climate Regime, COP3 and the Kyoto Protocol', UNU/IAS, Tokyo, June 1998; and Sebastian Obetthür and Hermann E. Ott (1999), *The Kyoto Protocol, International Climate Change Policy for the 21st Century*, Berlin.

2. Climate Change: The Historical Context

Natural changes in the climate are an age-old phenomenon. Throughout the course of the Earth's history there have been significant alterations in global temperatures, which have subsequently threatened the livelihood of species and civilisations. These changes in the climate have often coincided with historic social disruptions.

For instance, the French Revolution evolved from a background of social unrest and widespread famine caused, in part, by unusually cold temperatures between 1783 and 1789, which created severe crop failures. The main source of these extremely cold years was the volcanic eruptions in Iceland and Japan.

The Irish potato famine was also triggered by climate change: A mass migration from Scotland to Ireland occurred due to cold weather, causing a sharp increase in Ireland's population. This situation was aggravated by a potato plague, which ruined the harvest, caused by new climate conditions in Ireland and human error in depending on a single monoculture. Such events demonstrate how vulnerable societies are to small, yet sudden, climatic changes.[7]

Industrialisation, population growth and deforestation have led to a significant increase in global greenhouse gas emissions. Scientists predict that this may lead to a great number of adverse effects. Sea levels may rise as a result of melting glaciers, flooding coastal areas and low-lying islands in the coming decades. Precipitation levels may change in some regions, harvests may be threatened in many countries, and biological diversity may decline. Moreover, the risk of natural catastrophes, such as floods, droughts and storms, may increase in many places. Diseases may spread to new areas, water shortages may intensify and environmental migration may explode as a result.

This section investigates the development of the scientific understanding of climate change as well as the build-up of a science-politics interface on the issue.

---

7 Al Gore (1993), *Earth in the Balance, Ecology and the Human Spirit*, New York, pp. 58, 60-61, 69-70, 75.

2.1 UNDERSTANDING THE SCIENCE – THE FIRST STEPS

The first milestone in the science of climate change[8] was reached in 1827, when French mathematician Jean Baptiste Fourier discovered that the atmosphere has an effect on the temperature of the Earth's surface. He introduced the analogy of a greenhouse to describe how the atmosphere retains heat like a glasshouse, thus affecting the surface temperature.[9] In the 1860s, British scientist John Tyndall stipulated that the ice ages might have been caused by a reduction in the atmospheric amounts of water vapour and carbon dioxide.[10]

Swedish chemist Svante Arrhenius was the first to suggest in 1908 that human industrial activities, e.g., the burning of fossil fuels, could significantly affect the Earth's climate. He calculated how changes in carbon dioxide (CO_2) concentration affect the Earth's temperature: A doubling of CO_2 concentration in the atmosphere would lead to increases in the temperature by 5-6°C. These figures are remarkably close to current scientific predictions. However, rather than anticipating possible costs to future generations, he emphasised the positive effects from a more temperate climate for the northern regions.[11]

In 1931, a Japanese fairy tale, 'Gosukôburi no Deki', by Kenji Miyazawa,[12] was inspired by Arrhenius' work on the possibility of altering the climate by

8 There are two terms that refer to the issue at hand, *climate change* and *global warming*. Although they are often used interchangeably, there are some connotational differences between the two. *Climate change* is, strictly speaking, the more neutral term, as some regions are predicted to cool as a result of climate change. Climate change 'means a change of climate which is attributed directly or indirectly to human activity that alters the composition of the global atmosphere and which is in addition to natural climate variability observed over comparable time periods' (Article 1.2, Framework Convention of Climate Change (1992), UNEP/IUP, hereinafter abbreviated FCCC). *Global warming*, on the other hand, is a popular term for the enhanced greenhouse effect. It was banned from UN documents because it implied that a human-induced warming of the global climate was a proven fact, which not all governments had accepted. Brown (1996), *op. cit.*, pp. 13-14. This thesis mostly uses the more neutral term climate change.

9 Paterson (1996), *op. cit.*, p. 17.

10 Brown (1996), *op. cit.*, pp. 12-13.

11 Paterson (1996), *op. cit.*, pp. 19-21.

12 For reasons of simplicity, Japanese names will appear in the order first name followed by surname.

changing the concentration of greenhouse gases in the atmosphere.[13]

British meteorologist G. D. Callendar shared Arrhenius' conjecture that global warming was controllable and favourable, as it might enhance agricultural produce for a growing world population. However, in providing evidence in 1938 that average temperatures had already increased between the 1880s and 1930s, he substituted Arrhenius' theoretical assumption with real and compelling data. He thus concluded that human emissions of trace gases were sufficient to lead to significant climatic change.[14]

Despite a reversal of this warming trend during the period 1940-1970, in which mean temperatures dropped by a margin of 0.2°C, the scientific scope of meteorological research significantly broadened from the 1950s onward. The reasons for this were not purely scientific.

Technological developments after the Second World War created new demands, particularly in the US, for enhanced meteorological data. This secured an increase in investments in climate change science. The research was necessary to develop and refine jet aviation, satellites, computers and radio communications, as well as the nuclear bomb. It was facilitated by the establishment of the first permanent CO_2 monitoring station at Mauna Loa, Hawaii in 1957,[15] which has generated meteorological data for future large-scale research into the climate system.

Hence, during the period from 1827 to the 1940s, research on the climate system remained peripheral and relatively loose, and served purely scientific interests. The effects of a warming climate were generally regarded as

---

13 Research efforts into the Earth's climate system, however, only started in the 1970s in Japan. Miranda A. Schreurs (1996), 'Domestic Institutions, International Agendas, and Global Environmental Protection in Japan and Germany', Dissertation, University of Michigan, published by University Microfilms International, Ann Arbor, MI., p. 232.

14 Paterson (1996), *op. cit.*, p. 21. Brown (1996), *op. cit.*, p. 13.

15 The site proved convenient due to its geographical distance from local pollution sources. J. W. Anderson (1998), 'The Kyoto Protocol on Climate Change, Background, Unresolved Issues and Next Steps', Resources for the Future, January 1998, accessible via <http://www.rff.org/about_rff/anderson.htm>. Furthermore, 1957 was a crucial year in climate change research. In addition to the opening of the first CO_2 monitoring station in Hawaii, 1957-58 was designed as an 'International Geophysical Year'. The numerous meetings that were held in this year initiated a process for connecting scientists from different disciplines and nationalities into networks of collaboration and peer review. Jäger and O'Riordan (1996), *op. cit.*, p. 14.

favourable to humankind. It was not the phenomenon of climate change itself but technological innovation during the 1950s that promoted huge investments into meteorological science and launched the development of a formidable research infrastructure.

2.2 THE POLITICISATION OF CLIMATE CHANGE

From the 1950s onwards, institution-building[16] and a substantial number of international meetings took place, which collated evidence that human-induced emissions were already affecting the climate. An internationally-endorsed scientific consensus on climate change and its potential adverse effects was eventually reached. As this evidence became increasingly compelling, policy-makers made their entry into the debates and embarked upon the difficult venture of devising political measures and response strategies to combat the possible adverse effects of climate change.

Two additional developments impacted on the politicisation process: Firstly, there was the rise of public environmental awareness which was spurred by a number of extreme climate events, wide media attention in the industrialised countries and a growing number of environmental pressure groups. Secondly,

16 At that point, some form of institutionalisation had already been in place. The establishment in 1878 of the International Meteorological Organisation (IMO) had provided the framework for international cooperation and standardisation of meteorological observations; it was reestablished as the World Meteorological Organisation (WMO) in 1951 under the UN umbrella with a refined focus on scientific and technical issues. Other institutions include the World Climate Research Programme (WCRP), the International Geosphere-Biosphere Programme (IGBP), the International Human Dimensions Programme on Global Environmental Change (IHDP), the Global Climate Observing System (GCOS), the Global Terrestrial Observing System (GTOS), the Global Ocean Observing System (GOOS), the World Weather Watch (WWW) and the Global Atmospheric Watch (GAW). For an overview of these institutions and their activities see International Council for Scientific Unions (1996), *Understanding Our Planet*, 2nd ed., Paris.

there were the successful negotiations of the ozone treaty. They are expanded upon in Chapters 2.2.1 and 2.2.2.

2.2.1 The Build-up of Scientific Evidence

In the 1970s and 1980s, a series of significant scientific meetings was held. *Table 1* lists these meetings, showing the various numbers of individual and country participants, the delegate constituencies and the conclusions adopted at these meetings.

Table 1: International meetings

Year, Location **Title**	**No. of partici-pants / countries**	**Delegate constitu-ency**	**Conclusions**
1970, Williamstown, Mass., USA Study of Critical Environmental Problems	n.a.	n.a.	Stressed potential long-term consequences of rising CO_2-levels and called for enhanced research into the matter.
1971, Wijk, Sweden Study on Man's Impact on Climate	30 / 14	Scientists	Assembled possible climatic effects. Major background paper on climate change issues for Stockholm.
1972, Stockholm, Sweden UN Conference on the Human Environment	1200 / 114	Delegates	Established the United Nations Environment Programme (UNEP). Only vague reference to the possibility of human impacts resulting from climate change.
1979, Geneva, Switzerland First World Climate Conference	400 / 50	Scientists	Appealed to all governments to 'foresee and to prevent potential man-made changes in climate that might be adverse to the well-being of humanity'.
1985, Villach, Austria International Conference on the Assessment of the role of carbon dioxide and other greenhouse gases in climate variations and associated impacts	n.a.	Scientists	Consensus as regards to the fact that global mean surface temperatures increase by 1.5-4.5°C for a doubling of the atmospheric CO_2 concentration; called for research into policy options for responding to any potential climate change.

1987, Villach and Bellagio, Italy Developing Policies for Responding to Climate Change	n.a.	Scientists	Called for the examination of the need to move towards a convention.
1988, Toronto, Canada The Changing Atmosphere: Implications for Global Security	>300 / 48	Scientists, NGOs, IGOs, Governments	'Toronto Target': 20 per cent reduction of CO_2 emissions from 1988 levels by the year 2005, with the eventual aim of a 50 per cent cut.
1989, Noordwijk, The Netherlands Ministerial Conference on Atmospheric Pollution and Climatic Change	72 / n.a.	Ministers	Commitment to stabilising CO_2 emissions at levels to be set by the IPCC in its First Assessment Report.
1990, Geneva, Switzerland Second World Climate Conference	137 / n.a.	Scientists, Governments	*Scientific segment*: urged countries to take steps towards reducing sources and increasing sinks[17] of GHG emissions, and negotiate a global convention on climate change. *Ministerial segment*: proposed the adoption by governments of targets and/or national programmes or strategies limiting GHG emissions.
1992, Rio de Janeiro, Brazil UN Conference on Environment and Development	>1600 / 178	Delegates, NGOs, IGOs	The Framework Convention on Climate Change was open for signature.

Source: Paterson (1996), *op. cit.*, pp. 25-48; Brown (1996), *op. cit.*, p. 15, 39.

The first World Climate Conference in 1979 appealed to all governments to become active in the climate change process, and to 'foresee' and 'prevent' potential adverse effects from anthropogenic changes in the global climate. The topic thus took on a new political dimension.[18]

During the 1980s, this science-politics interface became increasingly pronounced: The Villach Conference of 1985, held in Austria, is often referred to as a break-through event in this respect.[19] Between 1987 and 1990, a number

[17] Sinks are ecosystems that absorb carbon dioxide, such as forests, soil and oceans.

[18] Brenton (1994), *op. cit.*, p. 165.

[19] Paterson (1996), *op. cit.*, pp. 28-29. Jäger and O'Riordan (1996), *op. cit.*, p. 17. Barrett and Chambers (1998), *op. cit.*, p. 3. Daniel Bodansky (1994), 'Prologue to the Climate

of crucial steps were taken to form an internationally recognised scientific consensus on the immediacy of climate change. In 1988, the 'Toronto Target' was set at a meeting of expert scientists and politicians in Toronto, Canada. It called for a 20 per cent reduction of CO_2 emissions of 1988 levels by the year 2005, with the eventual aim of reaching a 50 per cent cut. It was the first widely endorsed quantitative policy proposal.[20]

The 1990 Second World Climate Conference programme was split into a scientific and a ministerial segment. The deliberations from these two segments illustrate the divide between scientific and ministerial formulations, setting a precedent for future clashes between the two. The language of the former was clear and conclusive, while that of the latter was ambiguous and evasive. Whereas the scientists suggested the introduction of targets, ministers could only agree to 'national programmes and strategies (...) limiting emissions of greenhouse gases'. The use of this ambiguous term 'limiting' could be understood to mean less than reducing or even stabilising emissions – or it could simply imply reducing the rate of emission growth.[21]

Climate change thus became recognised not only as a scientific concern but also as a political and socio-economic one. This change in focus can be seen, for instance, in the structure of the Intergovernmental Panel on Climate Change (IPCC), set up by the United Nations Environment Programme (UNEP) and the World Meteorological Organisation (WMO) in 1988. The IPCC has a threefold mission:

- Working Group I (Science) is to assess the available scientific information on climate change;
- Working Group II (Impacts) is to assess the environmental and socio-economic impacts from climate change; and
- Working Group III (Responses) is to formulate response strategies.[22]

Change Convention', pp. 45-74, in: Irving M. Mintzer and J. Amber Leonard (eds.), *Negotiating Climate Change: The Inside Story of the Rio Convention*, Cambridge, p. 48.

20 Paterson (1996), *op. cit.*, pp. 33-34. Wendy E. Franz (1997), 'The Development of an International Agenda for Climate Change: Connecting Science to Policy', ENRP Discussion Paper E-97.07, Kennedy School of Government, Harvard University, Cambridge, Mass., pp. 25-28. Brenton (1994), *The Greening of Machiavelli, The Evolution of International Environmental Politics*, London, p. 166.

21 Paterson (1996), *op. cit.*, pp. 47-48.

22 Brown (1996), *op. cit.*, p. 18. Brenton (1994), *op. cit.*, pp. 168-169. Jäger and O'Riordan (1996), *op. cit.*, p. 16.

The IPCC's First Assessment Report, released in 1990, demonstrated an implacable divide between the three categories Science, Impacts and Responses. This is clearly portrayed in the results of the different working groups.

Working Group I concluded that in a 'business-as-usual' scenario temperatures would rise by 0.3°C per decade, which would be the fastest rate experienced within the past 10,000 years.[23] Although these findings were not significantly different from previous assessments, the IPCC Report was unique in two aspects: Firstly, it stated that increases in greenhouse gases (GHGs) 'will enhance the greenhouse effect, resulting on average in an additional warming of the Earth's surface'[24] - the first time such a decisive statement had been made. Secondly, its degree of representation was unprecedented and elevated the political status of its findings. It reflected the consensus of the world's 2,500 leading climate scientists, although they were primarily from the political North.[25]

Working Group II was less successful in providing definite and influential results. Whilst it agreed that the consequences of climate change for agriculture, wetlands, forests, coastlines and desertification could be substantial, and that adaptation costs could be high, it remained uncertain as to how the climate would change locally. Hence, the conclusions reached were not as clear and influential as those of Working Group I.

Working Group III became the centre for much of the climate change politics of the subsequent 18 months. It included in its conclusions one important recommendation: Policy-makers should begin to negotiate a global framework convention on climate change.[26]

23 J.T. Houghton, G.J. Jenkins and J.J. Ephraums (eds.) (1990), *Climate Change, The IPCC Scientific Assessment*, Cambridge, p. xi.

24 *Ibid.*.

25 Disputes remain within, but especially between, disciplines over the science of climate change. Geologists and palaeonto-climatologists are sceptical of the current IPCC predictions. According to their measurements and data, the climate of the past 10.000 years has been unusually warm and stable. Other theories contend that sunspots and the interaction between cosmic rays and clouds could play important roles in explaining changes in the climate. *The Economist*, 21 February 1998, pp. 101-102. *The Economist*, 1 April 1998, p. 81. *DIE ZEIT*, 28 November 1997.

2.2.2 The Build-up of Environmental Momentum

In addition to the development of scientific evidence, growing environmental momentum in the 1980s spurred political action on the climate change issue. This environmental momentum was enhanced by a growing public awareness in many industrialised countries, on the one hand, and the successful outcome of the ozone negotiations, on the other.

During the economic prosperity of the 1980s, a growing environmental awareness developed in many industrialised countries, which, in turn, forced green politics onto their political agendas.[27] This movement was nurtured by two major developments, which were both significantly covered by the media.

Firstly, in addition to a number of environmental disasters of international scope, such as Chernobyl, the Sandoz spill into the Rhine and the Exxon Valdez accident off the coast of Alaska, several extreme hurricanes and droughts took place in the mid- to late 1980s.[28] Furthermore, this decade was the hottest on record.[29] These events created an upsurge of environmental activity across the globe, fuelled by a growing sense of insecurity about the climate.[30] The media devoured these extreme climate events and played a pivotal role in shaping public opinion from a predominantly alarmist point of view.[31]

---

26 Brenton (1994), *op. cit.*, pp. 177-183. Paterson (1996), *op. cit.*, pp. 40-47.

27 Eugene B. Skolnikoff (1990), 'The Policy Gridlock on Global Warming', pp. 77-93, in: *Foreign Policy* 79, Summer 1990, p. 79. Yasuko Kawashima (1997), 'A Comprehensive Analysis of the Decision-making Process of Developed Countries toward CO_2 Emissions Reduction Targets', pp. 95-126, in: *International Environmental Affairs*, Vol. 9, No. 2, 1997, pp. 110-111.

28 They include a freak hurricane in the English Channel in 1987; Hurricane Gilbert, which caused widespread damage in the Caribbean in 1988; the breaking off of a 40x160 km large iceberg in the Antarctic, leading to fears of sea-level rise; and a catastrophic drought in the American Midwest in 1988; Brown (1996), *op. cit.*, p. 19.

29 Mathew Paterson and Michael Grubb (1992), 'The international politics of climate change', pp. 293-311, in: *International Affairs*, 68, 2(1992), p. 294.

30 Brenton (1994), *op. cit.*, p. 166. Brown (1996), *op. cit.*, p. 19.

31 Skolnikoff (1990), *op. cit.*, p. 87. For an examination of the news coverage of extreme weather events in the US see Sheldon Ungar (1999), 'Is Strange Weather in the Air? A Study of U.S: National Network News Coverage of Extreme Weather Events', pp. 133-150, in: *Climate Change*, Vol. 41, No. 2, February 1999; and in Japan see Schreurs (1996), *op. cit.*, pp. 240-241. See also Hans-Peter Peters and Marion Sippel (1998), 'Der Treibhauseffekt als journalistische Herausforderung', pp. 293-316, in: Peter Borsch and

Secondly, the international environmental movement, made up of a network of non-governmental organisations (NGOs) became increasingly active in this process. NGOs, such as Greenpeace, World Wildlife Fund and Friends of the Earth, provided and circulated information on environmental issues, lobbied governments at local, national and international levels, and consulted with various agencies of the United Nations system. Moreover, NGOs exerted substantial influence on public opinion through the media and various publicity stunts.[32]

The other factor that spurred the politicisation of the climate change issue was a major advancement in another environmental process: ozone depletion. The successful negotiation of the Montreal Protocol in 1987 and the publicity it attracted set a precedent for the climate change process that encouraged the prospect for similar international cooperation on other global environmental issues.[33]

The alarm about depletion of the ozone layer was raised in 1985, when the discovery of an 'ozone hole' over the Antarctic became public.[34] It was the first environmental problem that captured worldwide attention. It was a concept that was comprehensible to both journalists and the general public, and was immediate enough for policy makers to act upon. Within two years, a protocol - calling for a 50 per cent cut of the use of chlorofluorocarbons (CFCs) of 1986 levels (Article 2.4, Montreal Protocol) – was open for signatures.[35]

Importantly, in both the ozone and the climate cases, the first political agreements were in place before there was complete scientific consensus on the

---

Jürgen-Friedrich Hake (eds.), *Klimaschutz: eine globale Herausforderung*, Landsberg a.L., pp. 299-300.

32 Charles W. Jr. Kegley and Eugene R. Wittkopf (1997), *World Politics, Trends and Transformations*, 6th ed., New York, 145-148. Barrett and Chambers (1998), *op. cit.*, pp. 30-31.

33 For an excellent analysis of the ozone negotiations see Richard E. Benedick (1991), *Ozone Diplomacy*, Cambridge, Mass.. See also Thomas Gehring (1996), 'Das internationale Regime zum Schutz der Ozonschicht: Modell für das Klimaregime', pp. 49-60, in: Hans Günther Brauch (ed.), *Klimapolitik*, Berlin.

34 Benedick (1991), *op. cit.*, pp. 18-19.

35 The Montreal Protocol is included in Benedick (1991), *op. cit.*, pp. 230-241. Its commitments were further strengthened by the London Revision to the Montreal Protocol of 1990. See Table 13.1 in *ibid.*, pp. 190-195.

issues at stake.[36] The greatest difference between the two cases was one of scale. Whilst the phasing out of CFCs mainly affects the CFC industry and CFC-intensive products, emission reductions of GHGs will have an impact on the whole energy sector. This will have a major effect on the entire global economy[37] and make the problem of climate change very difficult to overcome.

Hence, by late 1990, a series of international conferences, greater public and media awareness, a strident environmental lobby, the first IPCC Report, and calls by a strong scientific community for political response, had created an unstoppable momentum towards political action. Negotiations for an international convention on climate change could not be delayed any further.

2.3 The Framework Convention on Climate Change

In December 1990, the United Nations General Assembly created the Intergovernmental Negotiating Committee on Climate Change (INC) (Resolution 45/212) to negotiate a convention text. After 16 months of heated negotiations the United Nations Framework Convention on Climate Change (UNFCCC) was completed. It was adopted at INC-9 on 9 May 1992 in New York and opened for signatures in Rio de Janeiro at the United Nations Conference of Environment and Development (UNCED), in June 1992. With regard to the climate issue, the Rio Summit merely provided the occasion for signing the Convention.[38] Although it is a legally binding document, it conveys through its ambiguous language a wide and irreconcilable array of the various players' interests, and the many scientific, economic and social uncertainties

36 Brown (1996), *op. cit.*, pp. 21-27. Oberthür and Ott (1999), p. 11.

37 Brenton (1994), *op. cit.*, p. 167.

38 For details on the negotiation process see Daniel Bodansky (1993), 'The United Nations Framework Convention on Climate Change: A Commentary', pp. 451-558, in: *The Yale Journal of International Law*, Vol. 18, No. 2, Summer 1993, pp. 481-492; and Brenton (1994), *op. cit*,, pp. 185-195.

permeating the issue of climate change. This sub-chapter analyses the objective, principles and commitments enshrined in the Convention.

2.3.1 Objective

The 'ultimate' objective[39] of the Convention is the 'stabilisation of greenhouse gas concentrations in the atmosphere at a level that would prevent dangerous anthropogenic interference with the climate system'. It should be 'achieved within a time-frame sufficient to allow ecosystems to adapt naturally to climate change, to ensure that food production is not threatened and to enable economic development to proceed in a sustainable manner' (Article 2, FCCC). The objective is referred to throughout the Convention text.[40]

However, it leaves ample room for interpretation. The objective determines that the emphasis is to be placed on global concentration rather than individual emissions, and leaves the level and time-frame of the objective unspecified.[41] It speaks of 'dangerous anthropogenic interference', which, according to ROWBOTHAM, is an undefined term in the Convention and is thus subject to varied interpretations by the parties.[42] Furthermore, different regions will experience different levels of impact, and will thus require individual time-frames for implementation. The Convention objective, as DOLZER asserts, is a result-oriented description rather than a date-oriented concept.[43]

---

39 The term 'ultimate' stands in clear contrast to 'immediate' or 'direct'. Rudolf Dolzer (1994), 'The Legal Status of the Commitments in the Convention on Climate Change and the Need for Future Revisions', pp. 131-145, in: Ted Hanisch (ed.), *Climate Change and the Agenda for Research*, Boulder, p. 134.

40 It appears in Article 2; Article 3; Article 4.2 (a), (d) and (e)(i); and Article 7.2 (a), (d), (e) and (m), FCCC.

41 Stefan Schuppert (1998), *Neue Steuerungselemente im Umweltvölkerrecht am Beispiel des Montrealer Protokolls und des Klimaschutzrahmenübereinkommens*, Max-Planck-Institut für ausländisches öffentliches Recht und Völkerrecht, Band 131, Berlin, p. 179.

42 Elizabeth J. Rowbotham (1996), 'Legal Obligation and Uncertainties in the Climate Change Convention', pp. 32-50, in: Jill Jäger and Tim O'Riordan (eds.), *Politics of Climate Change: A European Perspective*, London, pp. 34-35.

43 Dolzer (1994), *op. cit.*, pp. 134-135.

2.3.2 Principles

In reaching the 'ultimate objective' of the Convention, Article 3, FCCC, spells out five principles, which should guide the actions and measures undertaken by the parties. They are considered to be important normative elements of the Convention.[44]

The first one is the principle of common but differentiated responsibility (Article 3.1), which has been incorporated into all environmental treaties since the Rio Earth Summit.[45] It implies that the industrialised country parties should take the lead in combating climate change and its adverse effects as they bear the historical responsibility for this problem.

Secondly, the specific needs of developing countries should be given full consideration, especially of those countries that are particularly vulnerable to the adverse effects of climate change (Article 3.2).

Thirdly, the Article highlights the precautionary principle, which was first explicitly referred to in the 1985 Vienna Convention on Ozone Depleting Substances and is included in all Rio documents.[46] It implies that lack of full scientific certainty should not be a reason for postponing measures to counter climate change and its adverse effects. Because of the weakness of scientific evidence at the time, the precautionary principle was central to the Convention's composure. However, the principle is closely linked to an economic precondition that measures should be cost-effective (Article 3.3).[47]

As a fourth principle, the Article mentions sustainable development. It emphasises that policies and measures to protect the climate system should be commensurate with national circumstances and promote economic development because it constitutes an essential precondition for introducing measures to

---

44 Hermann Ott (1996), 'Völkerrechtliche Aspekte der Klimarahmenkonvention', pp. 61-74, in: Hans Günter Brauch (ed.), *Klimapolitik*, Berlin, p. 65.

45 Udo E. Simonis (1998), 'Das "Kyoto-Protokoll", Aufforderung zu einer innovativen Klimapolitik', Wissenschaftszentrum Berlin für Sozialforschung, FS II 98-403, Berlin, p. 8.

46 James Cameron (1999), 'The Precautionary Principle', pp. 239-269, in: Gary P. Sampson and W. Bradnee Chambers (eds.), *Trade, Environment, and the Millennium*, Tokyo, pp. 240-241.

47 Bodansky (1993), *op. cit.*, pp. 503-504.

combat climate change (Article 3.4).

Lastly, the Article invites parties to cooperate in promoting an open international economic system that would lead to sustainable economic growth, especially in developing countries (Article 3.5).[48]

2.3.3 Commitments

Article 4, FCCC, on commitments is the centrepiece of the Convention. The parties are committed by Article 4.2 (a) to take 'corresponding measures on the mitigation of climate change, by limiting its anthropogenic emissions of greenhouse gases, and protecting and enhancing its greenhouse gas sinks and reservoirs'. Furthermore, Article 4.2 (b) commits Annex I parties[49] to 'aim' to return 'individually or jointly to their 1990 levels [of] these anthropogenic emissions of carbon dioxide and other greenhouse gases not controlled by the Montreal Protocol'.

The language suggests significant latitude as Annex I parties are merely required to *aim* to achieve these commitments by means of undertaking relevant measures. Hence, if a party's final achievement falls short of the set target of returning emission levels to 1990 by the year 2000, it will not necessarily be in

48 *Ibid.*, p. 505.

49 The Convention text contains two *annexes* in order to differentiate between the various groups of parties, which are bound by different commitments under it. *Annex I* lists 24 OECD countries plus the European Community and 12 Central and Eastern European states with economies in transition (EIT). It does not include Mexico and the Republic of Korea, which joined the OECD in 1994 and 1996, respectively. Annex I parties are committed to adopting policies and measures aimed at reducing GHG emissions to 1990 levels by the year 2000. Croatia, Liechtenstein, Monaco and Slovenia have acceded to Annex I since ratification of the Convention, and the Czech Republic and Slovakia are now listed separately. Decision 4/CP.3, contained in document FCCC/1997/7/Add.1, which was adopted at COP3, amends Annex I of the Convention to incorporate these changes. *Annex II* lists all OECD member countries as of 1992, i.e., all Annex I countries minus EITs, which are to provide 'new and additional financial resources to meet the agreed full costs incurred by developing countries' and facilitate technology transfer (Article 4.3, FCCC). *Non-Annex I parties* are all other parties that are not included in Annex I of the Convention.

breach of its obligations under the Convention.[50]

Article 4.2 (d) requires the Conference of the Parties (COP) to 'review the adequacy of subparagraphs (a) and (b)' at its first session, and undertake a 'second review' by no later than 31 December 1998.[51]

Article 4.1 spells out further means to achieving the goal of the Convention. These constitute core commitments for *all* parties to the Convention. They include the following:

- Publication of national inventories of anthropogenic emissions of greenhouse gases (Article 4.1 (a));
- Implementation and publication of national programmes to address anthropogenic emissions of greenhouse gases (Article 4.1 (b));
- Promotion and cooperation in the development and diffusion of technologies and practices which reduce or prevent anthropogenic emissions of greenhouse gases (Article 4.1 (c));
- Consideration of climate change in social, economic and environmental policies (Article 4.1 (f));
- Promotion and cooperation in the exchange of information related to the climate system and to the economic and social consequences of response strategies (Article 4.1 (h));
- Promotion and cooperation in education, training and public awareness related to climate change (Article 4.1 (i));
- Communicate to the Conference of the Parties information related to implementation, in accordance with Article 12 (Article 4.1. (j)).

In this light, the Article should be regarded as a first step towards minimising emissions of GHGs, and in achieving the ultimate objective of the Convention.

---

50 Rowbotham (1996), *op. cit.*, p. 37. The in-depth review of national communications conducted by an independent team of reviewers under the UNFCCC Secretariat plays an important role in determining the implementation of commitments. Rowbotham (1996), *op. cit.*, pp. 44-45.

51 The second review of adequacy of commitments under the Convention was discussed at COP4 in Buenos Aires, Argentina, in November 1998. However, agreement was not reached. Grubb *et al.* (1999), p. 250. Agreement was still not reached at COP5. FCCC/CP/1999/6.

Figure 1: Institutional arrangements of the FCCC

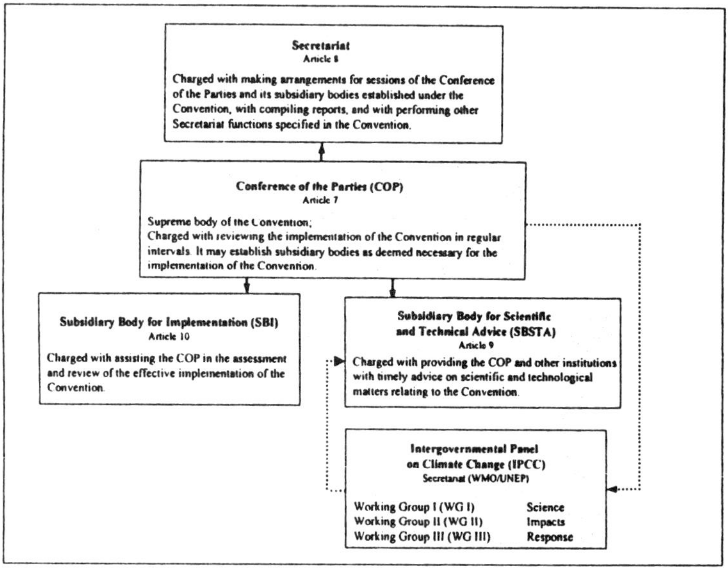

In summary, the Convention sets the institutional and procedural foundations for a process to establish the objective of the Convention. It requires all parties to inventory their GHG emissions and to develop mitigation and adaptation strategies, and it includes a non-binding reduction target. Moreover, the Convention mandates regular reviews of the adequacy of commitments and an annual COP to discuss further action and make decisions to improve the implementation of the Convention.

Through these provisions the Convention has a built-in ratchet effect,[52] which drives the FCCC process forward. The 'framework' nature of the Convention and its convoluted language reflect the status of scientific evidence, and the political limitations to the consensus possible at the time of its adoption.

The next chapter expands upon these political limits, and analyses the interests of the major players in greater detail.

---

52 Brown (1996), *op. cit.*, p. 46.

3. ACTOR ANALYSIS : THE EU, US AND JAPAN

States remain the decisive players in international deliberations despite growing influence from the wide array of non-governmental actors. The power of a state in the climate change process is determined by its level of GHG emissions,[53] by its economic capacity[54] and by its ability to shape the negotiation process. FERMANN asserts that 'Germany, Japan and the United States all belong to the rather exclusive, but heterogeneous group of *critical actors* capable of contributing substantially to solving or worsening the problem of global warming'.[55] Whilst Germany is a key player within the EU, the EU's 15 member states share a common negotiating position and speak with one voice in the international climate negotiations. This study thus focuses on the EU-level.

---

53 The EU accounts for 24.2 per cent, the US for 36.1 per cent and Japan for 8.5 per cent of total CO_2 emissions of Annex I parties in 1990; FCCC/CP/1997/7/Add.1. Combined, this is around two-thirds of all Annex I parties' emissions.

54 See, for example, 'Denver Summit Countries: Key Economic Data', June 1997, at: <http://www.state.gov/www/issues/economic/summit/fsg7data97.html>.

55 Gunnar Fermann (1997), 'Political Leadership and Climate Change: The Prospects of Germany, Japan and the United States', pp. 341-405, in: Gunnar Fermann (ed.), *Politics of Climate Change, Key Issues and Critical Actors*, Oslo, p. 343. There is a substantial selection of literature that analyses leadership roles in the FCCC process. See, for example, Gunnar Fermann (1994), 'Political Leadership and the Development of Problem-solving Capacity in the Global Greenhouse: Prospects for Germany, Japan and the United States Towards the 21st Century', EED Report 1994/3, The Fridtjof Nansen Institute, Lysaker; Fermann (1997), *op. cit.*; and Joyeeta Gupta and Nicolien van der Grijp (forthcoming) 'Leadership in the Climate Change Regime: The European Union in the Looking Glass', to be published in the *International Journal of Environment and Pollution*.

During COP3, the press reported primarily on the EU, US and Japan. See, for example, *International Herald Tribune*, 10 December 1997; *The Daily Yomiuri*, 9 December 1997; and *The Japan Times*, 11 December 1997.

For a theoretical analysis of power in negotiations see Francisco Javier Guerroero Aguirre (1995), 'Power, Asymmetry and Negotiation: A Theoretical Analysis', Summer 1995, at: <http://snipe.ukc.ac.uk/international/papers.dir/paco1.html> as of 10 January 1999.

This chapter analyses the Macro-Perspective (policies and positions) and the Micro-Perspective (interests and internal structures) of the EU, US and Japan in connection with the issue of climate change. Furthermore, it gives an overview of other important actors and alliances in the climate change process.

3.1 MACRO-PERSPECTIVE

This section examines the three actors' climate change policies and negotiating positions at COP3. Although the EU consists of 15 member states, each to a large degree serving its own national interest, it has continued to share a common negotiating position and speak with one voice in the climate change debate. Both the European Community and its individual member states are parties to the Convention. However, because the EU lacks a government body with relevant competencies, in this case in climate policy, the EC – represented through the European Commission – has not played a major role in the international climate negotiations.

The US and Japan have had similar negotiating positions on many issues. They are both members of an informal alliance called JUSSCANNZ (Japan, US, Switzerland, Canada, Australia, Norway and New Zealand). This alliance was formed in the early stages of the climate negotiations, mainly as a counterweight to the EU because of its commitment to stronger targets. JUSSCANNZ is a loose coalition of parties, which has often invited other parties to attend its meetings, such as Iceland, South Korea and Mexico. Each of these countries has had separate negotiating positions and strategies, with Switzerland often at odds with the rest of the group.[56] Therefore, this chapter investigates the US and Japanese climate change policies and positions separately.

[56] See also Barrett and Chambers (1998), *op. cit.*, p. 41; and Grubb *et al.* (1999), *op. cit.*, p. 34.

3.1.1 The European Union

3.1.1.1 Climate Change Policy

The climate change issue was taken up by EC[57] institutions as a response to political action at national and international levels.[58] Elicited by the 1985 Villach conference, and promoted primarily by Germany, the Netherlands and Denmark,[59] the issue was first raised in the European Parliament in 1986.[60] The Commission issued its first Communication on the subject in November 1988.

---

57 For reasons of simplicity, this study generally refers to the European Community (EC) when dealing with legal or policy issues, and to the European Union (EU) when dealing with political issues after its establishment in 1993. The European Union was established by the Maastricht Treaty in 1993 and rests on three pillars: common foreign and security policy, home affairs and justice policy, and the European Community. The European Community includes agricultural, transport and environment policies. It has the legal competency to sign and ratify international treaties, such as the Convention and the Kyoto Protocol; the European Union does not. Richard Macrory and Martin Hession (1996), 'The European Community and Climate Change, The role of law and legal competence', pp. 106-154, in: Jill Jäger and Tim O'Riordan (eds.), *Politics of Climate Change*, London, p. 142, n. 2.

58 Andrea Lenschow (1996), 'Der umweltpolitische Entscheidungsprozeß in der Europäischen Union am Beispiel der Klimapolitik', pp. 89-104, in: Hans Günter Brauch (ed.), *Klimapolitik*, Berlin, p. 91.

59 Wolfgang Fischer (1992), 'Die Klimapolitik der Staaten zwischen globaler Verantwortung und nationalen Interessen', pp. 164-223, in: P. Borsch and P.M. Wiedemann (eds.), *Was wird aus unserem Klima?*, Landsberg a.L., p. 210.

60 Despite the formidable expansion of action on the environment since 1972, the climate change issue received recognition by the EC only in the second half of the 1980s. Prior to this, some research on the issue had been conducted through a Commission-initiated research programme from 1979-1986. The Fourth Action Programme on the Environment of 1987, covering 1987-1992, had referred to climate change merely as a subject for further research. The Fifth Action Programme (1992-2000), however, does make reference to climate change. It states, for example, that climate change is a global problem and thus requires concerted action at the regional, as well as international, level. 'Fifth European Community environment programme: towards sustainability', at: <http://europa.eu.int/scadplus/leg/en/lvb/128062.htm> as of 4 May 1999. For a detailed account see Gerrit Vonkeman (1996) 'International co-operation: the European Union', pp. 105-134, in: Andrew Blowers and Pieter Glasbergen, *Environmental Policy in an International Context: Prospects for Environmental Change*, London.

The European Council – the highest EC body – called for early adoption of targets and strategies for limiting emissions of GHGs at its Dublin meeting in June 1990. This was followed, in October 1990, by a joint Council of Energy and Environmental Ministers Agreement to stabilise CO_2 emissions at 1990 levels in the EC as a whole by the year 2000. This collective target allowed substantial leeway for individual variation, taking into account a member country's energy efficiency and stage of economic development.[61] On the one hand, it catered to those countries that were eager to act on the EC level and were prepared to take on higher reduction targets, whilst, on the other hand, serving those that were willing to lend their support to the ambitious EC position but were not willing to take on measures to reduce emissions themselves.

Table 2: National CO_2 abatement targets for the EC member states (1990)

	Target	Mill. tons of CO_2	Per capita
Belgium	Reduction of 5% by 2000 at 1990 levels	112	11.2
Denmark	Reduction of 20% by 2005 at 1988 levels	53	9.9
France	Stabilisation at less than 2tC per capita	366	6.5
Germany	Reduction of 25% by 2005 at 1990 levels	1005	13.0
Greece	'Realistic objective' plus 15% between 1990 and 2000	74	7.4
Ireland	Increase limited to 20% by 2000 at 1990 levels	31	8.8
Italy	Stabilisation by 2000 at 1990 levels	402	6.9
Luxembourg	Stabilisation by 2000 at 1990 levels	12.5	35.0
Netherlands	Reduction of 3% by 2000 at 1990 levels	157	12.2
Portugal	Increase limited to 40% by 2000 at 1990 levels	40	4.1
Spain	Increase limited to 25% by 2000 at 1990 levels	211	5.4
UK	Stabilisation by 2000 at 1990 levels	579	10.2

Source: Collier (1997), *op. cit.*, p. 59; Loske (1996), *op. cit.*, p. 279

However, the Council established that this EC target was linked to an 'assumption' – which was rather an unspoken condition – that other leading countries would undertake similar commitments. PATERSON claims that by making the commitment dependent on like action elsewhere, the EC lost any

61 This common EC target glosses over the fact that most EC member countries had no independent national climate policies at the time. Fischer (1992), *op. cit.*, p. 210.

credibility it had had as a world leader on climate change.[62] Given the nature of the climate change issue, i.e., that reductions are only meaningful if implemented on a worldwide scale, it is paramount that as many countries as possible participate. Therefore, it can be argued that it was indeed an act of leadership from the EC to exert pressure on other countries to curb emissions too.

HAIGH/WAGNER maintain that the 1990 Agreement was a conscious attempt by the EC to emerge as a leader in the expanding field of international environmental politics.[63] The strong EC commitment, and the pressure it exerted on other countries, especially the US, contributed positively to the outcome of the Convention negotiations. Albeit not achieving the EC's dedicated proposal, the Convention does include specific targets and timetables, making its commitments more significant than they would have been otherwise.[64]

However, the downside to sharing common proposals and positions has been increased inflexibility and inability to respond to changing positions of negotiating opponents. One country speaks on behalf of all others, although 15 voices might give much more weight to their common position. Moreover, EU coordination has been time-consuming and has achieved only lowest common denominator proposals and positions, while neglecting negotiations with other parties.[65]

Despite these deficiencies, negotiators from EU member states have stressed the advantages arising from this scheme: 'If the Netherlands speaks with a few other countries, people may listen politely. But, if the EU speaks, people listen seriously. It is more powerful'.[66]

Following the Convention negotiations, the Commission made an ambitious

---

62 Paterson (1996), *op. cit.*, p. 88.

63 Nigel Haigh (1996), 'Climate Change Policies and Politics in the European Community', pp. 155-185, in: Jill Jäger and Tim O'Riordan (eds.), *Politics of Climate Change*, London, pp. 161-162. Wagner (1997), *op. cit.*, pp. 311-312.

64 Wagner (1997), *op. cit.*, p. 334. For a more detailed account on the nature of the EC in the negotiation process, see Sylvia Schumer (1996), 'Die Europäische Union als Akteur in der internationalen Umweltpolitik am Beispiel des Klimaregimes', pp. 105-114, in: Hans Günter Brauch (ed.), *Klimapolitik*, Berlin, pp. 109-110.

65 Personal communication with Joanna Depledge, 26 January 2000.

66 Norichika Kanie, 'Domestic Policy and Leadership in Multilateral Diplomacy: The Case of the Netherlands' Kyoto Protocol Negotiation', UNU/IAS Environment and Multilateral Diplomacy Series, No. 3, Tokyo, p. 11.

policy proposal in June 1992 on measures to reach the EC target. These included a directive on a carbon and energy tax, a directive on energy efficiency within the existing SAVE (Specific Actions for Vigorous Efficiency) programme, a decision on renewable energy within the EC's ALTENER programme, and a decision concerning a monitoring mechanism for CO_2 emissions. Except for the contentious directive on a carbon and energy tax, the measures were adopted as EC legislation in 1993.[67]

However, the implementation of these EC policies and measures proved difficult, mainly due to conflicting national interests and the seemingly low priority of the issue. CAMPBELL maintains that the SAVE directive proposal has become an ineffectual framework directive, and the ALTENER programme has suffered from a lack of resources and compliance.[68]

Furthermore, the combined carbon and energy tax was opposed by France, which, due to its strong reliance on nuclear energy, favoured a straight carbon tax. It had also been opposed by the less developed member states, which saw their economic development in jeopardy, and the UK, which objected to ceding taxation from national governments to the EC level. Eventually, Austria, Denmark, Finland, the Netherlands and Sweden implemented a carbon tax jointly among themselves.[69]

The only successful policy turned out to be the monitoring mechanism, which became the cornerstone of EC climate change policy. It was more precise about the nature of the national programmes than the Convention itself, and the EC targets were higher than those of the Convention. Whilst the responsibility to devise measures to achieve the set targets remained with the member states, the mechanism actually put pressure on them to meet their individual targets. The Commission's responsibility was to ensure that the national programmes added

67 Jay P. Wagner (1997), 'The Climate Change Policy of the European Union', pp. 297-340, in: Gunnar Fermann (ed.), *International Politics of Climate Change, Key Issues and Critical Actors*, Oslo, pp. 316-318. See also Joy Hyvarinen (1999), 'The European Community's Monitoring Mechanism for CO_2 and other Recent Developments', pp. 191-197, in: *Review of European Community & International Environmental Law*, Volume 8, Issue 2, 1999.

68 Karen Campbell (1998), 'From Rio to Kyoto: The Use of Voluntary Agreements to Implement the Climate Change Convention', pp. 159-169, in: *Review of European Community & International Environmental Law*, Volume 7, Issue 2, 1998, pp. 163-164.

69 *Ibid.*, p. 168, n. 50.

up to a stabilisation of emissions in the EC as a whole.[70] Hence, the mechanism provided the EC with a monitoring and enforcement tool, which was stricter than what had been negotiated at the international level.

The rather disappointing outcome of the EC's policies and measures introduced a somewhat less euphoric phase in climate change policy,[71] which was compounded by the reluctance of the US to discuss targets and timetables.[72] Yet, at the first Conference of the Parties (COP1), in 1995, the EU composed a counterweight to the laggard US and OPEC countries, and ultimately achieved, together with the developing countries[73], a compromise in the form of the 'Berlin Mandate'. It laid the groundwork for binding targets and timetables by aiming to set limitation objectives within specified time-frames by 1997.[74]

However, the surprise announcement by the US during COP2, in 1996, that it was willing to negotiate legally-binding targets and timetables, and the anticipation that the EU countries might be able to achieve a stabilisation of their GHG emissions by the year 2000,[75] created a new momentum for action. Yet, the

---

70 Haigh (1996), *op. cit.*, pp. 164-167.

71 Macrory and Hession (1996), *op. cit.*, p. 149. Ute Collier (1997), 'The EU and Climate Change Policy: The Struggle over Policy Competences', pp. 43-64, in: Ute Collier and Ragnar E. Löfstedt, *Cases in Climate Change Policy, Political Reality in the European Union*, London, p. 61.

72 Gupta and Grijp (forthcoming) *op. cit.*.

73 The developing countries participate in the climate change deliberations as the Group of 77 and China (G77 & China); China often associates itself with the positions of the G77. Their common concern is to ensure that the economic and social development of its members is not jeopardised by disproportionate efforts to tackle an environmental problem that was historically created by other countries. However, it comprises around 130 countries, which again are divided into various alliances because of their acutely conflicting interests. On the one hand, there is AOSIS that is particularly vulnerable to the adverse effects of climate change, primarily sea-level rise, and is thus keen on international action. On the other hand, it includes OPEC that is pointedly sensitive to the potential short-term problem of declining oil revenues resulting from abatement measures. The G77 & China have experienced difficulty in keeping common positions, which is mainly due to the fact that the group is extremely heterogeneous, and that its members are not trustful enough of each other to collaborate more effectively. See Yamin (1998), *op. cit.*, p. 366.

74 The Berlin Mandate is contained in document FCCC/CP/1995/7/Add.1, Decision 1/CP.1.

75 However, this would happen due to fortuitous circumstances rather than progressive policy. Merely two countries appeared to be able to overshoot their targets: It seemed

EU was still not able to push through internationally coordinated and legally binding policies and measures (PAMs) to be included in a future protocol, which it strongly endorsed in order to maintain its competitiveness vis-à-vis other industrialised countries.[76]

In summary, the record of the EC as a party to the Convention can be seen as varied. As a political actor, the EC has shown leadership in promoting climate change policy internationally. The EC member states have achieved more by taking on common positions and proposals, whilst struggling with conflicting internal interests and negotiating inflexibility. It has been successful at putting pressure on key countries for stronger commitments, and has thus managed to upgrade the commitments spelled out in the Convention. However, the outcome of its climate change policy still falls short of initial intentions. Its results have been modest, yet they are the aggregate of national policies and measures, and go beyond those laid down in the Convention.

3.1.1.2 Negotiating Position for COP3

The EU Environment Council adopted an ambitious negotiating position for COP3 in March 1997. It called for a legally binding commitment by industrialised countries to a flat emissions reduction rate of 15 per cent of 1990 levels by the year 2010. The EU envisaged achieving these reductions by an internal burden-sharing arrangement, thereby allowing the less developed EU countries to actually

possible that Germany would achieve it because it inherited dirty and inefficient coal plants from East Germany upon reunification, which it subsequently shut down to alleviate unhealthy air pollution and improve economic efficiency. Economic decline in the former East Germany has further contributed to Germany's emission reductions. The UK was believed to achieve substantial reductions because of Thatcher's decision to cut back on coal subsidies and switch from coal to natural gas, which was at least in part induced by her desire to weaken the power of the miners union. Eugene B. Skolnikoff (1997), 'Same Science, Differing Policies; The Saga of Global Climate Change', Report No. 22, MIT Joint Program on the Science and Policy of Global Change, Cambridge, Mass., August 1997, at: <http://www.mit.edu/globalchange/www/rpt22.html>.

76 Markus Ehrmann and Sebastian Oberthür (1997), 'Spring in Climate Negotiations?', pp. 192-196, in: *Environmental Policy and Law*, 27/3 (1997), p. 193.

increase their emissions.[77]

It did not, however, support differentiation outside the EU. It supported a basket of three greenhouse gases, consisting of carbon dioxide, methane and nitrous oxide. At the same time, it opposed inclusion of new commitments for developing countries in the protocol. The EU agreement did not represent a unilateral pledge – it meant to be binding only if other industrialised countries, most notably the US, agreed to similar binding reduction targets.[78] In June 1997, the EC environment ministers added to their proposal a preliminary 7.5 per cent reduction target of 1990 levels by 2005.

The inclusion of a set of binding common and coordinated policies and measures in the protocol was a key objective of the EU. It proposed creating three annexes of policies and measures (PAMs): one comprising common mandatory PAMs, a second listing PAMs to be accorded high priority, and a third listing PAMs that might be included in national programmes depending on national circumstances.[79] Such an approach would foster better implementation of the commitments under the Convention and additional protocols.[80]

The EU proposal was the outcome of a 7-month negotiating phase; after quite some juggling of national targets, an EC-wide 10 per cent target evolved. However, this was still regarded as too low to influence the positions of the US and Japan in the protocol negotiations. Hence, it was decided to maintain the 15 per cent target and regard the 10 per cent shared-out contributions as an interim solution, to be adjusted after the US and Japan had made similar pledges.[81]

Thus, the core elements to the EU position were strong commitments from all industrialised countries with mandatory policies and measures to implement them.

77 The emission reductions were to be distributed among the EU member states in the following way: Austria: -25%; Belgium: -10%; Denmark: -25%; Finland: +/-0%; France: +/-0%; Germany: -25%; Greece: +30%; Ireland: +15%; Italy: -7%; Luxembourg: -30%; Netherlands: -10%; Portugal: +40%; Spain: + 17%; Sweden: +5%: United Kingdom: -10%. Ehrmann and Oberthür (1997), *op. cit.*, p. 193.

78 Ehrmann and Oberthür (1997), *op. cit.*, p. 193.

79 Yamin (1998), *op. cit.*, p. 116. FCCC/AGBM/1997/2.

80 Yamin (1998), *op. cit.*, p. 117.

81 For a detailed account of the internal EU negotiations see Lasse Ringius (1997), 'Differentiation, Leaders and Fairness, Negotiating Climate Commitments in the European Community', CICERO Report 1997:8, Oslo, pp. 18-33.

3.1.2 The United States

3.1.2.1 Climate Change Policy

In the 1960s and 1970s, a generously funded US research community vigorously developed and proliferated scientific knowledge on climate change.[82] In the 1980s, it helped expand the issue of climate change beyond the scientific realm and into the international political arena.[83] The US had developed a leadership role in the ozone negotiations, and was keen to promote its newly found capacity in the climate change process as well. George Bush even incorporated the climate change issue in his 1988 presidential election campaign. He announced that he would 'confront the greenhouse effect with the White House Effect'.[84] Senator Timothy Wirth called a hearing in the same year, to which he invited a high-calibre expert, James E. Hansen, director of NASA's Institute for Space Studies. Hansen alleged that he was 99 per cent certain that the cause of the spectacularly hot temperatures during that summer was man-made and not due to natural variation.[85]

Although the scientific evidence about the consequences of climate change became more compelling, Bush exchanged his supportive stance to a hard-line position against climate change policy. He rejected binding targets and timetables to reduce CO_2 emissions,[86] opposed technology transfer and financial

---

82 For instance, US scientists published a paper in 1975 revealing that CO_2 was not the only gas with a greenhouse effect. In 1979, the US National Academy of Sciences published a report in which they concluded that a doubling of CO_2 concentration in the atmosphere would lead to an increase of the Earth's surface mean temperature of between 1.5 and 4.5°C. Jäger and O'Riordan (1996), *op. cit.*, p. 14. Brown (1996), *op. cit.*, p. 17.

83 National Research Council (1998), 'Overview Global Environmental Change: Research Pathways for the Next Decade', Washington, D.C., pp. 4-5. Fermann (1997), *op. cit.*, p. 347.

84 Gore (1993), *op. cit.*, p. 10. Michael Hatch (1993), 'Domestic Politics and International Negotiations: the Politics of Global Warming in the United States', pp. 1-39, in: *Journal of Environment and Development* 2, 2 (Summer 1993), p. 10.

85 The New York Times printed the story at the top of page one. Anderson (1998), *op. cit.*.

86 He opposed them because of the potential increase of US emissions due to changes in population growth, world fuel prices and economic growth, and because a binding commitment might require policy actions with unforeseen economic consequences. Paul

commitments to developing countries,[87] advocated a 'no-regrets' approach,[88] and refuted the IPCC's scientific assessments. The official rationale for the shift in policy was that the science was still too premature and uncertain to warrant expensive emission reduction schemes. However, economists anticipated huge domestic costs of adaptation and response measures.[89]

HARRIS and PATERSON/GRUBB indicate that the uncompromising tone of the Bush administration was rooted to a large extent in the personal antipathy to environmental concerns of White House chief of staff John Sununu, and some State Department officials appointed by him.[90] NITZE contends that Sununu believed that the climate change issue was being used by environmentalists to impose their anti-growth agenda on the US economy, whilst Bush simply failed to come to terms with the issue.[91]

After Sununu's departure from office in early 1992, US climate policy softened to some extent. It began to endorse the findings of the IPCC, but continued to underscore the need for further research. Whilst still stressing the principles of no regrets and voluntarism, and the need for flexibility and cost-

---

G. Harris (1998), 'Understanding America's Climate Change Policy: Realpolitik, Pluralism, and Ethical Norms', OCEES Research Paper No. 15, Oxford, p. 48. Apart from Turkey, the US was the only country at the time that had not committed itself to unilateral abatement targets and timetables. Fermann (1997), *op. cit.*, p. 386.

87 Bush resented new aid commitments to developing countries, especially at a time of economic recession. Brown (1996), *op. cit.*, p. 33.

88 This meant that the Bush administration was prepared to only adopt policies that would have beneficial effects even if climate change was not taking effect. Fermann (1997), *op. cit.*, p. 375.

89 The 1988 Environment Protection Agency Report, 'The Potential Effects of Climate Change on the United States', forecasted a shrinkage of US forests, floods along the coastline, major crop losses, increased air pollution and a rise in electricity demand which could require spending $100 billion on new power plants over the next two decades. Brenton (1994), *op. cit.*, p. 167. Furthermore, the US was openly criticised for misusing its chairmanship of IPCC's Working Group III to block the development of response strategies that would have far-reaching economic consequences. Brown (1996), *op. cit.*, p. 36.

90 Harris (1998), *op. cit.*, pp. 44-45. Paterson and Grubb (1992), *op. cit.*, pp. 305-306.

91 Nitze alleges this to have been due in large part to Bush's personal ignorance. William A. Nitze (1994), 'A Failure of Presidential Leadership', pp. 187-200, in: Irving M. Mintzer and J. Amber Leonard (eds.), *Negotiating Climate Change: The Inside Story of the Rio Convention*, Cambridge.

effectiveness in implementing measures, the US demanded the involvement of all countries in the process. Furthermore, it acknowledged the need for financial aid to developing countries to permit their participation in the Convention.[92]

The US signed and ratified the Convention in 1992,[93] which it might not have done had Sununu remained in office.[94] Although it secured the removal of the provision for legally-binding targets and timetables from the final version of the Convention,[95] it still agreed to a document which laid the foundations for further commitments. By acknowledging the Convention's objective and principles, the Bush administration accepted the built-in ratchet effect, and thus the irreversibility of a process toward ever-tighter commitments on national governments.

The Clinton administration, which took over the White House in 1993, was more sympathetic to the climate change issue. Two months into office, Clinton proposed a broad tax on all energy consumption. On Earth Day in April 1993, he announced voluntary targets and timetables and included them in his 'Climate Change Action Plan' later that year. The wording, however – to 'return US greenhouse gas emissions to 1990 levels by the year 2000 with cost effective domestic actions' – demonstrates that the Clinton administration did not part from Bush's original policy principles of voluntarism, cost-effectiveness and flexibility.[96] Moreover, prospects for further policies substantially narrowed after the Republicans won back both Houses of Congress after 1994.[97]

During COP1 in 1995, the US once again rejected legally-binding targets and

---

92 Harris (1998), *op. cit.*, p. 47.

93 The US was the fourth country to ratify the Convention on 15 October 1992.

94 Harris (1998), *op. cit.*, p. 44.

95 However, according to a study conducted by the US Environment Protection Agency (EPA), the current and planned environmental measures were already sufficient to stabilise GHG emissions at 1990 levels by 2000. Thorsten Görrissen (1993), *Grenzüberschreitende Umweltprobleme in der internationalen Politik*, Baden-Baden, p. 175.

96 Those measures which were not voluntary, such as the proposed energy tax, failed due to opposition from the industrial lobbies. Loske (1996), *op. cit.*, p. 267. Pieter van Beukering and Pier Vellinga (1996), 'Climate change: From science to global policies', pp. 187-215, in: Peter B. Sloep and Andrew Blowers (eds.), *Environmental Policy in an International Context: Environmental Problems as Conflicts of Interest*, London, p. 205. Campbell (1998), *op. cit.*, pp. 162-163.

97 FCCC/IDR.1/USA.

timetables. Yet, the final document, the 'Berlin Mandate', laid the groundwork for such binding targets by aiming to set limitation objectives within specified time-frames by 1997.[98]

The most significant turning point in US climate change policy occurred in 1996 at COP2. The US announced that it actively supported negotiations for a legally-binding international agreement by COP3 with realistic targets and timetables. However, it also demanded maximum national flexibility, including a tradable permit system.[99]

There are four main reasons for this drastic change in position. Firstly, the scientific evidence was more compelling by 1996.[100] Secondly, changes in personnel after Clinton's election to the White House led to greater negotiating flexibility. Thirdly, industrial opposition to political action on climate change lessened in this time (see Chapter 3.2.1). Fourthly, the administration identified a set of flexibility mechanisms, by which legally-binding reductions could be met without introducing unpopular 'command and control methods'.[101] Since 1996 was a presidential election year, refinement of the new US policy stance was postponed until after the elections.[102]

A direct response to the shift in the Clinton administration's climate change policy was Senate Resolution 98, sponsored by SenatorsByrd and Hagel, which passed 95-0 in July 1997. This so-called Byrd-Hagel Resolution made the congressional ratification of a protocol with binding targets and timetables conditional upon simultaneous binding commitments for developing countries.[103]

This resolution was intended to block the process towards achieving binding

98 Brown (1996), *op. cit.*, p. 168.

99 *Earth Negotiations Bulletin*, 22 July 1996.

100 The IPCC's Second Assessment Report concluded that there was 'a discernible human influence on global climate'. 'Summary for Policymakers: The Science of Climate Change – IPCC Working Group I' at: <http://ipcc.ch/pub/sarsum1.htm> as of 8 December 1999.

101 Campbell (1998), *op. cit.*, p. 162. Wolfgang Fischer and Petra Holtrup (1998a), 'Erfolge und Defizite der internationalen Klimaschutzpolitik', pp. 243-292, in: Peter Borsch and Jürgen-Friedrich Hake (eds.), *Klimaschutz: eine globale Herausforderung*, Landsberg a.L., p. 261.

102 Harris (1998), *op. cit.*, pp. 21-22.

103 Senate Resolution 98, 105th Congress, 1st Session, 1997, at: <http://www.senate/gov/~rpc/rva/1051/1051205.htm> as of 1 June 1999.

targets and timetables by inserting into the official US negotiating position a requirement that was knowingly unacceptable to a wide array of actors.[104] Furthermore, it was to deter short to medium-term distortion of competitiveness and its consequences from unilateral environmental regulations, and to enable US energy and high-tech companies to tap into the huge Chinese market.[105]

In summary, climate change policy in the United States has undergone a number of dramatic changes. Within a single decade it has shifted from the chief proponent of the science of climate change to the chief opponent of political response strategies. Only recently has the country moved to a more constructively tailored climate change policy, which includes a set of flexibility mechanisms by which it would be possible to achieve emission reductions voluntarily and cost-effectively. However, congressional demand for developing country commitments has introduced new constraints on US climate negotiators.

104 This can basically be explained by the Republican-led Senate's innate mistrust of any market-adverse, artificial standards 'imposed' by a developing country-dominated UN. Fischer and Holtrup (1998a), *op. cit.*, p. 259.

105 *The Japan Times*, 8 December 1997b.

3.1.2.2 Negotiating Position for COP3

On 22 October 1997, President Clinton announced his climate change proposal for COP3, containing three main elements.

Firstly, the US proposed a uniform binding target to reach 1990 emission levels by 2008-2012, and further reductions thereafter.[106] This target would include six GHGs (carbon dioxide, methane, nitrous oxide, hydrofluorocarbons, perfluorocarbons and sulphur hexafluoride). The US stressed that in contrast to the European proposal for more stringent early reductions its proposal was 'realistic', 'achievable' and 'meaningful'.[107] However, due to its sharp increase in GHG emissions in recent years, stabilisation at 1990 levels in the US already signifies an actual 23 per cent cut. A lower reduction would have therefore implied a nominal growth target, which would have been unacceptable to the international community. This proposal also included the concepts of 'borrowing' and 'banking'. In a multi-year emission budget, banking would enable emissions below the target in one period to be used to offset higher emissions in subsequent periods. Borrowing would allow parties to borrow against their targets for the next commitment period in order to emit more in the current period.[108]

Secondly, the proposal included a demand for developing country participation. It should be noted that the wording 'participation' is not as far-reaching as that of the Byrd-Hagel Resolution.[109]

---

106 Hence, it can be argued that the US consented to less than had been agreed upon in 1992. It thus went against the spirit of the Convention, which aimed to strengthen the commitments over time. On the other hand, it agreed to negotiate legally-binding targets and timetables, which carry decisively more weight than voluntary targets and timetables.

107 'President Clinton's Climate Change Proposal, 22 October 1997,' at: <http://www.state.gov/www/global/global_issues/climate/background.html> as of 30 October 1998.

108 Ehrmann and Oberthür (1997), *op. cit.*, pp. 193-194.

109 The Resolution reads that unless the protocol 'mandates new specific scheduled commitments to limit or reduce greenhouse gases for Developing Country Parties within the same compliance period' the Senate will not ratify it. 'Senate Resolution 98, 105th Congress, 1st Session, *op. cit.*.

It can be argued that this is justified on the grounds that it had not been evident during the Victorian industrial revolution that CO_2 would cause such harmful effects on the

Thirdly, in line with its preference for market, rather than policy-based instruments, the US proposed the establishment of an emissions trading system. This would consist of two flexibility mechanisms: international emissions trading and joint implementation.[110] The argument behind the adoption of flexibility mechanisms was that it did not matter to the environment where emission reductions were achieved. Therefore they should be met where the greatest gains could be attained at the least cost. The US argued that once a profit motive for enterprises to invest is created, the necessary forces to solve the climate change problem would be unleashed. The US is already experienced with pollution trading. The joint US-Canadian tradable permit system to control sulphur dioxide was widely regarded as a success as emissions were reduced at much lower cost than originally predicted.[111]

Thus, the primary objective of the US was to insert in the protocol as much implementation flexibility as possible.

3.1.3 Japan

3.1.3.1 Climate Change Policy

Japan was one of the first countries to actively encourage energy conservation. In the 1960s and 1970s, it controlled its CO_2 emissions, and reached stabilisation between 1973 and 1986. This was achieved whilst at the same time experiencing an average GNP growth rate of over 4 per cent.[112]

---

global climate and thus humankind. In contrast, today's industrialising countries are knowledgeable of these potentially devastating side-effects, and thus act less morally if they choose to retain their right to development via the same dirty and inefficient path. More justified, however, is the argument that the industrialised countries, which bear the historic responsibility for the problem and have the necessary financial and technological resources to reduce emissions, should take the first meaningful step in the process, as is established by the Convention.

110 'President Clinton's Climate Change Proposal, 22 October 1997', *op. cit.*.

111 *Financial Times*, 10 December 1997. Grubb *et al.* (1999), *op. cit.*, p. 90.

112 Gunnar Fermann (1993), 'Japan's 1990 Climate Policy under Pressure', pp. 287-300, in: *Security Dialogue*, Vol. 24(3), September 1993, p. 287. For a detailed account see

These early achievements in energy conservation resulted from three developments: Firstly, increasing competition from newly industrialised Asian countries in the late 1960s and early 1970s in labour-intensive manufacturing induced a major reorientation of investment. Japan moved away from heavy and energy-intensive industries such as shipbuilding, automobiles and machinery to high-technology industries such as microchips, consumer electronics and bio-technology. This shift had the additional effects of curbing demand for crude oil, as well as reducing CO_2 emissions.[113]

Secondly, in reaction to the oil crises of the 1970s, the Japanese government implemented a series of energy conservation and substitution measures. These included taxation to encourage private-sector investment in energy-efficient technology and measures to initiate energy substitution.[114]

Thirdly, after long-fought battles in the 1950s and 1960s by local environmentalists through the courts and the media to compensate the so-called Yokkaichi, Minamata and Itai-itai victims[115] from growing air and water pollution,[116] policy was implemented a decade later that resulted in further progress in energy conservation, and decreased CO_2 emissions.[117]

However, this impressive record of early energy efficiency measures was, to a

Fukasaku (1995), 'Energy and environment policy integration', pp. 1063-1076, in: *Energy Policy*, 1995, Vol. 23, No. 12, pp. 1066-1069.

113 Fermann (1993), *op. cit.*, p. 289.

114 *Ibid.*, pp. 289-290. Measures included a Sunshine Programme (1974) to develop technologies for new energy sources and a Moonlight Programme (1978) to develop energy conservation technologies. Takao Tamitate (1993), 'Japan's Climate Policy, A Rejoinder', pp. 301-304, in: *Security Dialogue*, Vol. 24(3), September 1993, pp. 302-303.

115 The Yokkaichi asthma was caused by noxious clouds of smog from industrial and automobile exhaust clouding over the cities. The Minamata and Itai-itai diseases were both caused by toxic compounds (methyl-mercury and cadmium, respectively) that industrial plants had discharged into the water. Weidner (1989), *op. cit.*, p. 481.

116 For more detailed accounts see Jun Ui (1986), 'Anti-Pollution Movement and Other Grass-Roots Organizations', pp. 109-119, in: Shigeto Tsuru and Helmut Weidner (eds.), *Environmental Policy in Japan*, Berlin, pp. 109-11; and David Wallace (1995), *Environmental Policy and Industrial Innovation, Strategies in Europe, the United States and Japan*, London, pp. 94-96.

117 Policies included switching to natural gas and nuclear energy. Natural gas increased from 1.6 per cent in 1973 to 9.6 per cent in 1986, and nuclear energy increased from 0.7 per cent in 1973 to 11 per cent in 1986. Fermann (1993), *op. cit.*, p. 290.

great extent, rolled back during the economic boom of the late 1980s. Business pressure, falling oil prices after 1986, and possibly lax policy enforcement reversed many of these efforts.[118] As a result, between 1987 and 1990, CO_2 emissions increased by 30 per cent.[119]

Japan entered the international climate change debate after several EC countries had criticised it for its 'wait and see'[120] posture on the issue at the 1989 Noordwijk Conference. Japan realised that it had underestimated the growing importance of environmental issues in international affairs.[121] Having learnt a lesson from its belated entry into the ozone negotiations,[122] Japan adapted to this new situation with great speed.

Thus, following a policy shift in the governing Liberal Democratic Party (LDP), Prime Minister Uno established a Ministerial Council on Global Environmental Protection in July 1989. It was to facilitate intra-governmental coordination of policies for negotiation at the international level. This Council played a central role in devising MITI's 'Action Programme to Arrest Global Warming', announced in October 1990,[123] which committed Japan to stabilising CO_2 emissions at 1990 levels by the year 2000 on a per-capita basis. This was in fact the first time in Japan's post-World War II history that it had taken a position different from the US.[124]

However, FERMANN points out that Japan's target was somewhat misleading and extremely weak. After a 30 per cent increase in CO_2 emissions from 1987 to 1990, and by choosing a per-capita linked stabilisation target, Japan allowed

---

118 Edward J. Lincoln (1993), *Japan's New Global Role*, Washington, D.C., p. 149.

119 Fermann (1997), *op. cit.*, p. 383.

120 *Ibid.*, pp. 375-376.

121 *Ibid.*, p. 287.

122 Schreurs (1996), *op. cit.*, p. 244.

123 *Ibid.*, pp. 243-244. For a translation of the Action Programme see Appendix 1 in Sozaburo Okamatsu (1992), 'MITI's Centennial Vision of Global Environment and Technology and the Response to Global Warming: Concerning New Earth 21', pp. 335-348, in: Helmar Krupp, *Energy Politics and Schumpeter Dynamics*, Tokyo, p. 344. The 1990 Action Programme was adopted through a government decision but not made into law. It stated that the government was reserving the option not to meet its target if other OECD countries failed to make sufficient efforts themselves. In the 1993 Basic Environmental Law, the Japanese Diet made no reference to abatement targets. Fermann (1997), *op. cit.*, pp. 384-385.

124 Beukering and Vellinga (1996), *op. cit.*, p. 208.

itself a de facto 40 per cent increase in CO_2 emissions between 1987 and 2000, and a 6 per cent increase between 1990 and 2000.[125]

One reason for Japan's failure to adopt a more assertive climate change policy was internal friction in its fractured environmental administration.[126] On the one hand, industrialists were concerned about the anticipated high abatement costs, especially since Japan was not expected to be badly hit by the adverse effects of climate change. Climate change policy might thus unnecessarily endanger economic growth.[127] On the other hand, Japan feared the indirect adverse impacts of a 'policy of neglect'[128] because climate change was a topic on the international political agenda.

The climate change issue provided its share of diplomatic gains. It constituted an opportunity for Japan to become more active in a global setting and enhance its international standing.[129] This was rated very highly at a time when Japan's environmental record was detracted by international criticism for its whaling practices, its massive consumption of tropical timber and drift-net fishing fleet.[130]

Furthermore, climate change was a non-military issue, providing an opportunity to build better relations with developing countries and thus enhancing Japan's economic security.[131] Most importantly, it was regarded as a potential for economic gain from growing exports of new, energy-efficient technologies and products.[132] FERMANN characterises Japan's incentives to cooperate in global climate change politics as notions of 'impact-scepticism' and

---

125 According to Fermann, the programme represented 'a political symbol, rather than an imperative'. Fermann (1997), *op. cit.*, pp. 383-384. It was the only compromise possible, which was mainly due to opposition by the Ministry of International Trade and Industry (MITI). Schreurs (1996), *op. cit.*, pp. 244-245.

126 15 ministries and agencies were involved in the deliberations. Fermann (1993), *op. cit.*, pp. 291-293.

127 *Ibid.*, p. 293. For a detailed analysis of climate change impacts on Japan see S. Nishioka and H. Harasawa (1998), *Global Warming, The Potential Impact on Japan*, Tokyo.

128 Fermann (1993), *op. cit.*, p. 292.

129 Neill Edward Harrison (1994), 'Heads in the Clouds, Feet in the Sand: Multilateral Policy Coordination in Global Environmental Issues', Dissertation, University of Denver, published by University Microfilms International, Ann Arbor, MI., p. 283-284.

130 *Der Spiegel* 25/1992.

131 Lincoln (1993), *op. cit.*, pp. 148-150.

132 *Ibid.*, p. 150.

'techno-optimism'.[133]

Thus, at the 1992 UNCED, Prime Minister Miyazawa[134] announced that Japan was prepared to play a leading role in the battle against pollution, and stabilise its carbon dioxide emissions at 1990 levels by the year 2000. In addition, he promised a significant amount of bilateral and multilateral aid for the environment to be spent in the coming five years.[135]

At COP2, Japan announced its invitation to hold COP3 in Kyoto. Its official reason for hosting COP3 was that it was eager to play a crucial role in the negotiations and wished to take on greater international responsibility. It would be the biggest international conference Japan has hosted to date. Furthermore, it aimed to extend its leadership role in the Asia-Pacific region.[136] Another major reason for this was Japan's aspiration to become a permanent member of the UN Security Council, and that its chances might improve by hosting a UN COP.[137]

In summary, Japan has been a rather low-profile player in the climate change issue, finding a middle ground between the more divergent positions of the EU and the US. Despite a good record of domestic energy conservation and efficiency improvement measures during the 1960s to mid-'80s, and advanced technological know-how, its position on climate change has been inconsistent. On the one hand, Japan's insistence that it was already very energy efficient on a per-capita base has impeded motivation to adopt more decisive commitments. On the other hand, aspirations to take on an active role in the international negotiations, enhance its foreign relations and build on its leadership capacity in the Asia-Pacific region encouraged it to strive ahead. By offering to host COP3, Japan has signalled good intentions to move the Kyoto process to a successful outcome.

3.1.3.2 Negotiating Position for COP3

On 7 October 1997, Japan announced its climate change proposal for COP3. It

---

[133] Fermann (1993), *op. cit.*, p. 292.

[134] He was the only major head of state not to attend the high-level segment of the Rio Summit. Lincoln (1993), *op. cit.*, p. 155.

[135] Schreurs (1996), *op. cit.*, p. 261.

[136] Petra A. Plate (1997), 'Japan und das Kohlendioxydproblem vor der Klimakonferenz in Kyoto', pp. 587-593, in: *Japan*, December 1997, p. 593.

[137] *ECO*, 1 December 1997. *Frankfurter Allgemeine Zeitung*, 1 December 1997.

should be regarded not solely as reflecting its national interests, but as a package seeking to combine the various interests and perspectives of the other parties in order to achieve a compromise proposal.[138] The Japanese proposal sets a maximum reduction target of 5 per cent of 1990 levels over the period 2008-2012, and comprises three GHGs (carbon dioxide, nitrous oxide and methane). However, this 5 per cent figure is somewhat misleading.

It includes a differentiation clause, by which the figure is substantially lowered depending on a country's emissions per GDP, emissions per capita and population growth.[139] By these standards, Japan would effectively only have to reduce its emissions by 2.5 per cent, and the Annex I countries as a whole by 3.2 per cent. In addition, the proposal included a flexibility clause, stipulating that countries, which miss their differentiated target by 2 per cent or less, would not be legally in breach of their obligations. Both devices added together significantly weaken a country's obligation. In effect, they reduce the commitment targets to 0.5 per cent and 1.2 per cent for Japan and all Annex I countries, respectively.[140] In addition, only the stabilisation of emissions was meant to be legally binding.[141] Furthermore, Japan suggested that emissions trading and joint implementation should only be adopted under certain conditions.

It is conceivable that the Japanese proposal for COP3 would have been weaker if it had been discussed with the US prior to its announcement. The Ministry for International Trade and Industry (MITI) had already negotiated a proposal with the industrial lobbies, which envisaged stabilisation of GHG emissions at 1990 levels, as further reductions were deemed unmanageable considering the economic situation.[142] Prime Minister Hashimoto's decision to

---

138 Udo E. Simonis (1997), 'Kyoto-Konferenz: Ein Rückblick auf die japanische Position zur internationalen Klimapolitik', pp. 452-456, in: *Zeitschrift für Angewandte Umweltforschung*, Jahrgang 10 (1997), Heft 4, p. 453.

139 While the Environment Agency lobbied for a flat-rate reduction, MITI succeeded in pushing through its preference for a per-capita based proposal. Richard Nathan (1997), 'Japan seeks to bridge split on emissions policy', p. 541, in: *Nature*, Vol. 387, No. 6634, 12 June 1997 p. 541.

140 'Japan: Positive Leadership from COP3 Chair Required!, ECO Tokyo Special Edition, November 8-9, 1997', at: <http://www.igc.org/climate/ets_jprop.html> as of 24 June 1999.

141 Ott (1998), *op. cit.*.

142 Plate (1997), *op. cit.*, p. 589. *Japan*, December 1997, p. 558.

announce Japan's proposal without prior negotiation with the key ministries – which by Japanese standards is very unusual[143] – might be explained by his aspiration to play a mediating role at COP3. Hashimoto had put forth a compromise proposal, which he was prepared to modify.[144]

Thus, the substance of Japan's proposal lay in its elaborate differentiation scheme. Otherwise, Japan held a middle position between the EU and the US because it was eager to reach consensus in Kyoto, no matter what the outcome would be. It preferred a meaningless agreement to no agreement.[145]

3.2 A MICRO-PERSPECTIVE

This sub-chapter will analyse and compare some of the different and conflicting interests and structures within the EU, US and Japan, which, to various extents, have shaped the national climate change policies and strategies discussed in the previous section. In this respect, it will begin with an analysis of industrial interests as, arguably, they are the most influential forces shaping national priorities. It will then scrutinise the institutional and political frameworks of the three actors, identify differences in their political and institutional interactions, and compare the channels through which industrial interests are communicated to decision makers. Finally, it will take into consideration the different energy supply structures, and discuss their levels of influence on the decision-making process.

---

143 *Japan*, December 1997, p. 558.

144 Petra A. Plate (1998), 'Ergebnisse der Klimakonferenz in Kyoto: Die Pflicht zum Forschen und Mitdenken', pp. 73-78, in: *JAPAN aktuell*, February 1998, p. 73.

145 Plate (1997), *op. cit.*, p. 593.

3.2.1 Industrial Interests

Climate change has an impact not only on the energy sector, but in fact on all industrial sectors. They have some common denominators, such as political, social and market stability, institutional legitimacy, as well as security. As climate change generates risk, volatility and vulnerability, it poses serious threats to industrial interests. In turn, climate change policy implies industrial and structural change.

Some industries are adamantly opposed to recognising the problem, as they fear that their interests might be threatened by climate change policy. Other industries, however, have a vested interest in climate change policy. These include not only sectors that are threatened to lose if the climate change problem is not recognised, but also sectors that seek to draw advantages from industrial and structural change. Whilst the 'losing' industrial sectors have developed effective access to the policy process through their powerful, well-organised and well-funded lobbies, as well as through representing large numbers of constituents, the potential 'winners' have not yet become as vocal and assertive.

'Losers' from climate change policy: The fossil fuel and energy intensive industries, including their respective labour unions, regard themselves to be the 'losers' from climate change policy options such as energy taxes and energy-efficiency incentives. These industries have argued that increases in electricity prices through energy taxes would increase production costs and result in a loss of competitiveness. This would cause job losses, and eventually drive businesses abroad where production costs are relatively cheaper.[146]

In the *European Union*, these industrial sectors have been primarily opposed to a carbon/energy tax as well as minimum energy-efficiency standards. The Union of Industrial and Employers Confederation (UNICE) and the European Petroleum Industry Association (EUROPIA) have lobbied against such measures, both at national and EU levels.[147]

However, European industry has not opposed the EU's commitment to enhance action on climate change at the international level, enabling the EU to hold a strong position throughout the climate negotiations. The only industrial concern was over the EU's competitiveness vis-à-vis other industrialised

---

146 Harris (1998), *op. cit.*, p. 51.

147 Wagner (1997), *op. cit.*, p. 328.

countries. This was why the EU insisted on enshrining mandatory policies and measures in the Kyoto Protocol and preferred a flat reduction rate for all industrialised countries.

In the *United States*, opposition by industrial lobbies to political action on climate change has been much fiercer than in the EU, both at national and international levels. Moreover, the methods used by American pressure groups are much bolder than in the EU. At the national level, several attempts to improve US climate change policy have been blocked through intense lobbying of Congress.[148] Tighter energy-efficiency requirements for American cars or the introduction of an energy tax have been prevented in this way.

In a bold attempt to gain public support for their interests, the powerful fossil fuel industry spent an estimated $13 million on a television, newspaper and radio advertising campaign between September and November 1997. Its message was that a Kyoto Protocol composed of legally-binding targets and timetables only for industrialised countries would entail serious economic costs for the US. Therefore, they argued that developing countries should participate in the commitments.[149]

At the international level, industrial lobbies have also acquired rather aggressive methods of pushing through their interests. They have established foundations and think tanks with green-sounding names, such as the 'Climate Council' or the 'Global Climate Coalition' (GCC), thereby disguising their pro-business and anti-green agendas.[150] The GCC has lobbied and shadowed the US delegation at climate conferences and passed on to Republican-led Congress anything opposing their interests, thus exerting effective pressure on the delegation.

The Climate Council was created by lawyer Donald H. Pearlman for the

148 Paterson (1997), *op. cit.*, p. 87.

149 Sharon Beder (1999), 'Corporate Hijacking of the Greenhouse Debate', pp. 119-122, in: *The Ecologist*, Vol. 29, No. 2, March/April 1999. Harris (1998), *op. cit.*, p. 50.

150 Richard N. L. Andrews (1997), 'United States', pp. 25-43, in: Martin Jänicke and Helmut Weidner (eds.), *National Environmental Policies, A Comparative Study of Capacity-Building*, Berlin p. 33. The GCC's web site can be found at: <http://www.worldcorpcom/dc-online/gcc/>. The GCC is a coalition of the fossil fuel sector, automobile manufacturing, agricultural firms and labour unions. Harris (1998), *op. cit.*, p. 49. The 'global' in its name is hardly justified: Out of 82 GCC representatives at FCCC conferences during 1996-97 merely two were non-US based. FCCC, *Who is Who in the UNFCCC Process 1996-1997*, 2nd Volume, Bonn.

purpose of becoming accredited to the Convention and thus being able to attend the climate conferences. For years, Pearlman has worked toward stalling the climate change process for his many anonymous industry customers. His aggressive methods have ranged from using Kuwaiti delegates as his mouthpiece to seriously obscuring the scientific integrity of IPCC meetings.[151] In this light, the US climate change policy stance of no regrets, voluntary action and deferral of measures to a time of greater scientific certainty reflects the interests of US fossil fuel and energy intensive industries.

In *Japan*, the industrial mouthpiece, MITI, has asserted only limited opposition to climate change policy. The automobile manufacturing and electronics sectors have voiced some concern, which was neither aggressive nor blatant, and remained more or less undetectable outside ministerial circles.[152]

'Winners' from climate change policy: There are at least three groups of industrial sectors that stand to win from more effective climate change policy.

Firstly, the insurance firms are exposed to huge potential costs if the climate change problem is not recognised. Insurance companies 'pay' in the future, and therefore the concept of pricing becomes more difficult proportionately to the greater risk of uncertainty due to a change in weather patterns. In addition, increasing public expectations of severe weather patterns provide the insurance sector with new business opportunities.[153]

Secondly, emerging industries such as energy technology and renewable energy industries profit from new demand and government incentives.

Thirdly, some of the traditionally 'losing' industries, such as the automobile and fossil fuel industries, are gradually and selectively acquiring a comparative advantage to their competitors by embarking early on the adaptation of their products to changes in policy and demand.

It should be noted that industrial support from the second and third group for climate action, and their adaptation to environmentally-sound technology, is driven to a large extent by aspirations for image improvement through green

151 *Der Spiegel* 14/1995, pp. 36-38. Paterson (1996), *op. cit.*, p. 81.

152 Schreurs (1996), *op. cit.*, p. 246.

153 Over 70 leading insurance companies from 25 countries have joined forces under UNEP auspices to adapt to changing circumstances. The 'Insurance Industry Initiative for the Environment' aims to enhance dialogue with shareholders, develop specialist investment strategies and sponsor projects such as the 'Solar Century'. Special Event, COP3, 5 December 1997, personal notes.

advertising.[154]

No major industry federation in *Europe* is formally opposed to the Kyoto Protocol.[155] Instead, a number of industries and corporations have announced impressive commitments. By committing itself to voluntarily reducing CO_2 emissions by up to 20 per cent of 1987 levels by 2005, German industry has provided the most ambitious example of climate-related measures within the EU.[156] Furthermore, Royal Dutch-Shell has established renewable energy as one of five core businesses, and announced in 1997 that it aimed to reduce corporate emissions by 10 per cent by 2002.[157]

In the *United States*, some industrial sectors are also gradually identifying potential gains from a restructuring of the energy mix. California's progressive anti-pollution policies have induced General Motors, Ford and other automobile manufacturers to develop pollution-free cars for the Californian market. Moreover, the large oil and gas company Enron, for example, is investing significantly in the solar market and building solarphotovoltaic power plants to generate electricity at a price competitive with fossil fuel.[158]

Given *Japan's* comparative advantage in energy-efficient technology, its industry is expecting internationally agreed-upon emission reduction commitments to boost technology exports.[159] Moreover, Keidanren[160] has for a long time adopted a supportive position. In 1991, it announced its own voluntary

154 Schreurs (1996), *op. cit.*, pp. 267-268.

155 Grubb *et al.* (1999), *op. cit.*, p. 258.

156 The textile industry, for example, achieved the highest target because it was almost completely relocated abroad. The damage done to the environment can be considered as worse than it would have been had conditions not been made so harsh in Germany. The ethics of voluntary commitments unfold well in Germany because moral pressure, as well as mutual trust and controls, have long been successful means to ensuring broad and committed participation. However, the aim of this pledge was not so much to counter climate change but to save on resources for future generations. Special Event, COP3, 9 December 1997, personal notes.

157 Grubb *et al.* (1999), *op. cit.*, p. 258.

158 Brown (1996), *op. cit.*, p. 191.

159 Kawashima (1997), *op. cit.*, p. 107.

160 Keidanren, the Japan Federation of Economic Organisations, is Japan's largest business association. It represents over 900 businesses and 120 business associations, including leading Japanese enterprises and foreign companies operating in Japan, thus making it Japan's most influential industrial organisation. See also Keidanren's web page at: <http://www.keidanren.or.jp/english/policy/pol058/index.html>.

programme to address climate change.[161] In September 1997, Keidanren issued a climate change policy paper, indicating its support for a protocol including flexibility and market mechanisms, but at the same time stressing its opposition to any kind of carbon or energy tax.[162]

In summary, industry in all three regions is opposed to energy taxes. While European industry has been concerned over competitiveness vis-à-vis the US and Japan, US industry has expressed fears over competition from key developing countries. Japanese industry is in favour of binding commitments because of enhanced technology export opportunities. Moreover, there is increasing support from within industry for more substantive climate action, even if the incentives are not always climate related.[163]

3.2.2 Governmental Structures

Governmental structures are integral to the process of foreign policy making. They provide distinct channels that determine what influence different actors within a state can exert on policy makers. Governmental structures affect the responsiveness of decision makers and their ability to change policy on a certain issue. The degree of openness of the political system determines, to a large extent, the influence of the general public and non-governmental actors, such as NGOs or the media, on the decision-making process. This section analyses how the governmental structures of the EU, US and Japan have shaped their climate change positions and policies.

The *European Community* institutions began to deal with the climate change

161 Anne K. Johnson (1998), 'The Influence of Political Culture on the Formation of Pre-Regime Climate Change Policies in Sweden, Japan and the United States', pp. 233-244, in: *Environmental Values*, Vol. 7, No. 2, 1998, p. 236.

162 'Japan Environment Quarterly, Vol.2 No.4, November 1997', at: <http://www.eic.or.jp/eanet/e/jeq/v002-04.html> as of 9 July 1999.

163 This development has gained speed since COP3. See, for example, Donald A. Carr and William L. Thomas (1998) 'The Kyoto Protocol and US Climate Change Policy: Implications for American Industry', pp. 191-201, in: *Review of European Community & International Environmental Law*, Volume 7, Issue 2, 1998.

issue at a time when there were no formal treaty provisions.[164] The 1993 Maastricht Treaty added a new dimension to Community policy; it sought to 'promote measures at international level to deal with regional or worldwide environmental problems' (Article 130r).[165] This reflects the increasing importance of global cooperation on matters such as climate change.

Despite considerable improvement in the powers of the EC in climate-related matters since the 1957 Rome Treaty, voting procedures and the subsidiarity principle[166] remain impediments to an effective EC climate change policy. Decisions over energy issues touch upon fiscal policy and thus require unanimity. As the EU member states have widely divergent interests in the issue, the development of a far-reaching common policy has not yet been possible.

The EC also lacks strong powers to implement, monitor and enforce legislation, and depends largely on its member states to do so.[167] Thus, due to

---

164 The 1957 Treaty of Rome provides merely indirect references to possible common environmental policy. Article 2 of the Treaty outlines in a broad manner the Community's tasks. These are to include 'a continued and balanced expansion' and 'an accelerated raising of the standard of living'. This was subsequently interpreted to include actions to safeguard environmental standards. Article 235 of the Treaty was interpreted to empower the Community to take whatever actions necessary to achieve environmental objectives. Derek Urwin (1991), *The Community of Europe, A History of European Integration since 1945*, Harlow. Wallace (1995), *op. cit.*, p. 232. The 1987 Single European Act (SEA) devotes Title XVI, Article 130r-t, to the Community's environment policy, and goes far beyond the simple economic requirements of the single internal market: Article 130r proclaims that the Community's environmental action should seek to preserve, protect and improve the quality of the environment, contribute to the protection of human health and ensure a prudent and rational utilisation of natural resources. Furthermore, Article 130s stipulates that action that includes fiscal policy or concerns the mix of energy sources of the member states is to be decided upon unanimously by the Council of Ministers, and by qualified majority voting otherwise. Dennis Swann (1992), *The Economics of the Common Market*, 7th ed., London, p. 330. Wagner (1997), *op. cit.*, p. 331.

165 'European Union – Selected instruments taken from Treaties, Title XVI, Environment', at: <http://europa.eu.int/abc/obj/treaties/en/entr6d16.htm#142> as of 4 May 1999.

166 The subsidiarity principle stipulates that the Community shall take relevant action to the extent at which the objectives can be attained better at Community level than at the level of the individual member states (Article 3(b), EC Treaty). Lenschow (1996), *op. cit.*, pp. 102-103.

167 Wagner (1997), *op. cit.*, pp. 329-331.

the intergovernmental nature of this process and the low EC competency in this field, the procedure of developing a common EC position is extremely time-consuming.[168] This explains to a large extent the disappointing outcome of the EU's common climate policy approach.

In the *United States*, both President and Congress are involved in the foreign policy making process. The US Constitution obliges the President to seek 'advice and consent' (Article II, Section 2) from the Senate for ratification of international treaties. It requires a two-thirds majority in the Senate, thus giving the Senate substantial powers in foreign policy.[169] Given the current Republican majority in the Senate, and the disparity between the Senate's position conveyed in its Byrd-Hagel Resolution of July 1997 and the contents of the Kyoto Protocol, ratification of the Protocol will be difficult.

Furthermore, the House of Representatives has power over the budget. The administration develops response strategies, and the House approves their funding. In recent years, it has approved merely 50 per cent of the administration's requests for funding of climate-related measures.[170] Moreover, since members of Congress represent states and regions, they usually deem those interests more important than foreign policy goals. In addition, state governments have important climate change policy levers over federal highway funds, building standards, speed limits and the electric utility industry.[171]

The openness and transparency of the political system in the US allows industry and public interests to exert considerable influence and control over the decision-making process.[172] Hence, the party dimension, with its different

168 Wolfgang Fischer and Petra Holtrup (1998b), 'Institutionelle Strukturen und Entscheidungsprozesse der Umweltaußenpolitik', pp. 121-136, in: Wolf-Dieter Eberwein and Karl Kaiser (eds.), *Deutschlands neue Außenpolitik, Band 4: Institutionen und Ressourcen*, Munich, p. 134.

169 Theodore J. Lowi and Benjamin Ginsberg (1998), *American Government*, Brief Fifth Edition, New York, p. 123. Konrad von Moltke and Atiq Rahman (1996), 'External Perspectives on Climate Change, A view from the United States and the Third World', pp. 330-345, in: Jill Jäger and Tim O'Riordan (eds.), *Politics of Climate Change*, London, p. 332.

170 FCCC/IDR.1/USA.

171 *Ibid.*

172 In addition to industry, scientists and economists have had a substantial impact on shaping both the Bush and Clinton administrations' climate change policies, arguing both for and against the issue. While some 'epistemic communities' – networks of

stakeholders emphasis, is a crucial determinant of the US position on climate change. Traditionally, Republican administrations are closely aligned to business and industry[173] and favour unrestricted capitalism. In contrast, Democrat-led administrations are close to trade unions and environmental groups, tending to promote a more equitable distribution of national wealth, and often not shunning from a more regulatory approach to environmental policy.[174]

Thus, impediments to a more affirmative position on climate change have come from the separation of powers in foreign policy making and from the openness of the government system, allowing for substantial influence of industrial interests.

Japan is a country that places great emphasis on avoiding confrontational diplomacy. It is generally 'not a challenger, but a supporter'.[175] Moreover, it has a long tradition of top-down organisation with a highly hierarchical power structure. It conducts preliminary negotiations to involve all actors, including the lower ranks, in order to reach consensus and prevent open conflict. This is commonly referred to as 'nemawashi'.

The Japanese policy-making process is extremely closed, making it difficult for stakeholder groups to influence politicians.[176] Competition between political actors is generally confined to the big ministries and the Liberal Democratic Party (LDP), rather than occurring between political parties or the executive and legislature.[177] However, the climate change issue has evoked some

knowledge-based experts, helping states identify their interests and proposing specific policies – have lobbied to change government policy on climate change, geophysicists close to the energy lobbies have provided research results that disproved the IPCC findings. Peter Haas (1992), 'Introduction: Epistemic Communities and International Policy Coordination', pp. 1-35, in: *International Organization* 46,1 (1992), p. 3.

173 75 per cent of $40 million donated by energy-related industries went to the Republican Party. *ECO*, 11 November 1998.

174 Johnson (1998), *op. cit.*, p. 240.

175 Takashi Inoguchi (1986), 'Japan's Images and Options: Not a Challenger, but a Supporter', pp. 95-119, in: *Journal of Japanese Studies*, Vol. 12, No. 1, Winter 1986, p. 95.

176 Johnson (1998), *op. cit.*, pp. 233-234.

177 The LDP has been in power from its formation in 1955 until today, with only short interruptions, and often controlled both the Lower and Upper Houses of Parliament. R. Vernon (1993), 'Behind the scenes: how policymaking in the EU, Japan and US affect global negotiations, pp. 12-20, 35-42, in: *Environment* 35(5), p. 19.

interdepartmental disputes: While MITI and the Ministry of Construction[178] see their interests jeopardised by too stringent laws, the Ministries of Foreign Affairs, Transport and Agriculture, the LDP and the Environment Agency argue for a constructive engagement in the climate negotiations.[179]

Despite domestic activism in environmental matters, there has been limited development of national movements devoted to preserving the global environment, such as the Greens in Germany or the Sierra Club in the US.[180] Compared with NGOs in Europe and the US, Japanese NGOs are considered weak and lacking in influence.[181]

In conclusion, articulation of the industrial standpoint to decision makers differs remarkably between the three regions. In the EU, the channels remain rather opaque because it is not a state – although it has developed many of its features. US industrial lobbies represent their interests through various channels, in which public opinion plays an important role. In Japan, industrial interests have well-established but intransparent direct channels to decision-makers. Industrial interests are well represented through MITI, making radical opposition from outside governmental circles redundant.

3.2.3 Energy Supply Factors

Climate change policy is intrinsically linked to energy policy. Therefore, factors, such as a country's indigenous resources and its energy mix, have a substantial impact on its climate change policy. The differences in the energy supply structures of the EU, US and Japan are outlined below.

Energy supply structures in the *European Union* reflect its inability to move beyond current policies, on the one hand, and its support for international measures, on the other. A key reason for the EU's poor record in devising an effective common climate policy are the vast structural differences between the

178 It feared that the new concern over the environment could translate into a decision to pass an environmental impact assessment law. Schreurs (1996), *op. cit.*, p. 245.

179 *Ibid.*, pp. 244-245, 249-250.

180 Vernon (1993), *op. cit.*, p. 20.

181 *The Daily Yomiuri*, 3 December 1997.

member states: While some countries rely heavily on nuclear energy (France and Belgium), others are extremely dependent on fossil fuels (Denmark and Greece). There are energy exporters (the Netherlands: gas; the UK: oil) and importers (primarily Denmark). Some countries have a large industrial sector (Germany), and others a relatively small one (Denmark).[182] Furthermore, CO_2 emissions vary considerably. While per capita CO_2 emissions are as low as 0.75 tons in Portugal, they amount to 6.46 tons in Luxembourg and 3.28 tons in Germany.[183]

However, since most EU countries are highly dependent on imported energy, the EU has supported strong international action and commitments on climate change. The simultaneous improvement of their balance of payments and reduction of dependence on OPEC countries have constituted welcome synergy effects to curbing climate change.[184]

Table 3: Energy consumption in the EU member states (1996), US and Japan (both 1991)

	Coal	Oil	Natural gas	Nuclear energy	Others
Austria	12%	50%	24%	14%	--
Belgium	18%	39%	22%	19%	2%
Denmark	35%	48%	17%	--	--
Finland	12%	44%	13%	22%	9%
France	4%	39%	12%	45%	--
Germany	27%	42%	18%	13%	--
Greece	34%	63%	3%	--	--
Ireland	23%	53%	21%	3%	--
Italy	6%	59%	28%	7%	--
Luxembourg	12%	57%	16%	15%	--
Netherlands	10%	39%	45%	3%	3%
Portugal	16%	79%	--	--	5%
Spain	15%	61%	7%	14%	3%
Sweden	5%	35%	2%	42%	16%
UK	16%	40%	30%	12%	2%
US	24%	40%	23%	8%	5%
Japan	17%	58%	10%	12%	2%

Source: Collier (1997), *op. cit.*, p. 46; Fermann (1997), *op. cit.*, p. 368

182 Loske (1996), *op. cit.*, p. 282.

183 Wagner (1997), *op. cit.*, p. 324.

The *United States* has huge indigenous energy resources, notably in Alaska. US coal reserves are estimated to last for another 400 years. Therefore, it is generally not deemed necessary in the US to seek new energy sources.[185] It is the biggest coal producer, providing one fourth of the world's coal. This comparative advantage has led to a specialisation in energy-intensive industry. Because of its large reserves of cheap coal it would be more costly for the US to substitute coal with another form of energy than it would be, for instance, for Germany. Germany is a relatively low emitter of CO_2 per unit of energy and currently subsidises coal production; it could thus save a lot of money by switching to gas.[186]

Japan is more energy-efficient than either the US or the EU, resulting from its high dependency on fossil fuel imports; only 2 per cent of its fossil energy supply are indigenous.[187] Due to this dependency – evident especially since the oil crises of the 1970s – Japan has developed superior energy conservation technologies and has expanded its use of nuclear energy.[188]

However, the increase in energy efficiency that has been accomplished in the past now works against Japan's achievements in climate policy insofar as that the marginal costs of current efforts in energy conservation are much higher.[189] Therefore, rather than committing itself to an ambitious reduction target, Japan prefers to see international recognition for past conservation achievements through its elaborate differentiation scheme, and extend international transfer of technology.[190]

Thus, while the constraints in the EU for a more stringent climate change

---

184 Paterson (1996), *op. cit.*, pp. 77-78.

185 Kawashima (1997), *op. cit.*, p. 108.

186 Bodansky (1993), *op. cit.*, p. 478.

187 Paterson (1996), *op. cit.*, p. 79.

188 Fermann states that nuclear power was assumed to increase from 12.3 per cent in 1990 to 16.6 per cent in 2000. However, the Japanese 'Nuclear Energy Programme' is already delayed due to the difficulty of finding appropriate sites in a densely populated country prone to volcanic eruptions and earthquakes. Fermann (1997), *op. cit.*, p. 370. For a more recent analysis on the status of nuclear energy in Japan see Yasuko Kawashima (2000), 'Nuclear Power and Climate Change: The Current Situation in Japan and a Message to the United States', Weathervane, Perspectives on Politics, February 2000, at: <http://www.weathervane.rff.org/pop/pop9/kawashima.html>.

189 Fermann (1997), *op. cit.*, p. 366.

190 *Ibid.*, p. 373.

policy approach lie in the vast differences in national energy structures, their overall dependence on imported energy provides for an incentive to adopt energy-conservation measures. The US, in contrast, has large indigenousenergy reserves that make it difficult to justify a policy shift away from fossil fuels without additional incentives. Japan's energy supply structure has, in turn, already made Japan very energy efficient, which makes reducing energy use further more costly.

4. THE KYOTO NEGOTIATIONS

The legal roots of the Kyoto Protocol lie in the review provisions of Article 4.2 (d), FCCC. This Article mandates the Conference of the Parties at its first session (COP1) to review the adequacy of the commitments of Annex I parties, which are included in Article 4.2 (a) and (b), FCCC. The review was to be carried out in the light of the best available scientific knowledge on climate change and assessment of its adverse impacts.

The first part of this chapter sketches the pre-Kyoto developments, i.e., COP1 and COP2, whilst the second part analyses the main issues, the negotiating process and the outcome of the Kyoto Conference (COP3).

4.1 THE FCCC PROCESS 1995-1997

4.1.1 COP1

The first session of the COP was held in Berlin, 28 March - 7 April 1995. As the IPCC's Second Assessment Report had not yet been finalised, COP1 lacked the bite of new and compelling scientific data on the gravity of the issue. Although the parties had agreed before COP1 that commitments under Article 4.2 (a) and (b), FCCC, were inadequate, there had been no agreement to change the Convention itself.[191] Thus, despite submission of a draft protocol by Trinidad and Tobago on behalf of AOSIS, containing binding targets and timetables

191 Timm Krägenow (1995), *Verhandlungspoker um Klimaschutz*, Freiburg/Br., p. 70.

similar to the Toronto Target,[192] as well as a document containing elements of a protocol submitted by Germany,[193] resistance from the US and OPEC to binding commitments blocked a more substantive outcome of COP1.

In the final hours of the conference, the COP adopted the Berlin Mandate,[194] which acknowledged that current commitments were inadequate to achieve the Convention's objective. The Berlin Mandate declares the intent of the parties to begin a process toward strengthening the commitments of Annex I parties beyond the year 2000 by means of adopting a protocol or another legal instrument.[195]

The addition 'of another legal instrument' was inserted on account of the US, and reflected its determination to keep the objective of the Mandate as open and flexible as possible. With the same intention, it also managed to omit mention of a base year from the document.[196] The COP established an Ad Hoc Group on the Berlin Mandate (AGBM) to begin the negotiation process for a protocol or another legal instrument. Furthermore, the Berlin Mandate ruled out any new commitments for developing countries.[197]

One of the issues that remained unresolved was joint implementation (JI). According to Article 4.2 (a) of the Convention, 'Parties may implement such policies and measures jointly with other Parties and may assist other Parties in contributing to the achievement of the objective of the Convention'.[198] It thus permits one country to help another to implement a project, or change a policy, so that the mutual sum of greenhouse gas emissions is lower than it would be otherwise.

The advantage to countries such as the US was that it offered a more efficient means of reducing global emissions and that it offered market economies the

---

192 The only difference to the Toronto Target was that the draft protocol reductions were to be set against a 1990 baseline, rather than the original 1988 baseline.

193 Documents A/AC.237/L.23 and A/AC.237/L.23/Add.1 respectively.

194 FCCC/CP/1995/7/Add.1, Decision 1/CP.1. For an elaborate analysis of the Berlin Mandate see Krägenow (1995), *op. cit.*, pp. 79-86.

195 In this thesis 'protocol' in small letters refers to 'a protocol or another legal document', while 'Protocol' refers to the Kyoto Protocol.

196 Krägenow (1995), *op. cit.*, p. 79.

197 Jäger and O'Riordan (1996), *op. cit.*, pp. 21-25. Paterson (1996), *op. cit.*, pp. 68-69. *Earth Negotiations Bulletin*, 10 April 1995.

198 In addition, this concept appears in Article 3.3 and Article 4.2 (d), FCCC.

chance to implement emission controls where they were cheapest.[199] The developing countries, on the other hand, feared that JI would become a means by which the rich countries could forego substantive commitments and structural adjustments. Furthermore, it could enable them to double-count their foreign aid flows.[200]

As a compromise, a pilot phase for 'activities implemented jointly' (AIJ) was set which would not involve any emissions credits for donor nation until 2000. The parties agreed that it should be introduced for the purpose of testing its practicality and suitable verification procedures.[201]

A further reason why COP1 did not achieve more ambitious results was that the political and economic conditions in 1995 were more arduous than in 1992. Recession, unemployment and excessive national public expenditure deficits beleaguered most governments much more than in 1992. Falling world energy prices made their impact felt on the Berlin negotiations. Furthermore, media attention in the industrialised countries to the issue of climate change had dropped after the Rio conference.[202]

The AGBM began its negotiations for a protocol in August 1995. The US presented itself as an assiduous laggard in stressing the importance of considering a myriad of research conclusions by numerous organisations before embarking on actual negotiations. At AGBM-2, the EU presented a possible structure for the protocol, including specified reduction targets for the years 2005, 2010 and 2020, as well as three long annexes of policies and measures. At AGBM-3, the final meeting before COP2, several EU member countries proposed specific reduction commitments. Germany, for example, proposed a 10

---

199 This would comply with Article 3.3, FCCC. Jäger and O'Riordan (1996), *op. cit.*, p. 21. Michael Grubb (1993), 'United Nations Framework Convention on Climate Change', pp. 61-73, in: Michael Grubb *et al.*, *The Earth Summit Agreements, A Guide and Assessment*, London, p. 66.

200 Jäger and O'Riordan (1996), *op. cit.*, pp. 23-24. Peter Hayes (1993), 'Constructing a global greenhouse regime', pp. 319-356, in: Peter Hayes and Kirk Smith (eds.), *The global greenhouse regime, Who pays?*, Tokyo, p. 322.

201 FCCC/CP/1995/7/Add.1, Decision 5/CP.1. *Earth Negotiations Bulletin*, 10 April 1995. Most projects under the AIJ scheme have been projects on energy efficiency, forest preservation and reforestation, and renewable energy. FCCC/SBSTA/1998/INF.3. For a history of the negotiations on AIJ, a list of relevant documents and updates on the methodological work on AIJ see: <http://www.unfccc.de/fccc/ccinfo/defaij.htm>.

202 Jäger and O'Riordan (1996), *op. cit.*, p. 27. Brown (1996), *op. cit.*, pp. 146-147.

per cent reduction target of 1990 levels by the year 2005, and further 10-15 per cent reductions by 2010.[203]

4.1.2 COP2

The second COP, held in Geneva, 8-19 July 1996, produced the 'Geneva Ministerial Declaration'.[204] The IPCC's Second Assessment Report (SAR), which had not been ready in time for COP1, was circulated in late 1995. Its conclusions were much more compelling than those of the First Assessment Report. The 'Summary for Policy Makers' pointed out that 'the balance of evidence suggests a discernible human influence on global climate', and projected global warming of between 1-3.5°C over the coming century.[205] The Geneva Declaration endorsed these findings, although the parties remained divided over the scope of the agreement.

The Declaration went beyond the Berlin Mandate in that it explicitly included the words 'legally binding', and called for greater expediency in negotiating legally binding targets, as stipulated in the Berlin Mandate, in time for COP3.[206] The surprise announcement by the US of its support for a protocol with legally binding targets and timetables, and thus its explicit inclusion in the document, had positive repercussions on the negotiation process. However, the US linked its support to a preference for maximum national flexibility and a tradable permit system, thus raising many new and complex questions for the negotiating sessions to follow.[207]

---

203 Ott (1997), *op. cit.*, p. 213.

204 The Geneva Declaration is included in document FCCC/CP/1996/15/Add.1.

205 'Summary for Policymakers: The Science of Climate Change – IPCC Working Group I', *op. cit.*. The Summaries are the politically crucial parts of the voluminous IPCC Assessment Reports as they are the only parts of the Reports that are read by non-scientists. Their contents are agreed upon line by line by governments.

206 Ott (1997), *op. cit.*, p. 213.

207 For a summary see *Earth Negotiations Bulletin*, 22 July 1996. It is interesting to note that the change in the US position evoked a change in the strategy of OPEC. Their income being highly dependent upon revenues from the production and export of fossil fuels, they put forth a proposal for a compensation mechanism. This mechanism was to

The Geneva Ministerial Declaration, however, was not adopted as an official conference document because Russia, Australia and the OPEC countries had objected to it. Yet, the Declaration marked a turning point upon which the negotiations in the final phase towards COP3 slowly built momentum.[208] The three AGBM meetings between COP2 and COP3 focused on preparing and streamlining the protocol negotiation text.

4.2 THE KYOTO CONFERENCE (COP3)

COP3 was held from 1-11 December 1997[209] in Kyoto, Japan. It was attended by over 10,000 people and was thus the largest climate change conference to date. At its final session on 11 December the COP adopted the Kyoto Protocol. It will come into force ninety days after ratification by at least 55 parties, incorporating Annex I parties which together account for at least 55 per cent of the total carbon dioxide emissions for 1990 of the parties included in Annex I of the Convention (Article 25).[210] The threshold for entry into force was set relatively high in order to ensure that the most important emitters would be

---

compensate developing countries for the adverse economic impacts of response measures by developed countries. Thus, instead of trying to obstruct the process altogether, the OPEC countries now fought to include in the protocol their specific interests. See Ehrmann and Oberthür (1997), *op. cit.*, p. 194.

208 For more details on these developments see Grubb *et al.* (1999), *op. cit.*, pp. 55-60.

209 The official conference deadline was 10 December but negotiations went into the following day. The dates were subsequently changed to 1-11 December 1997.

210 The Kyoto Protocol to the Convention on Climate Change (1998), UNEP/IUC, hereinafter abbreviated KP. As of 13 January 2000, 84 parties have signed and 22 have ratified the Kyoto Protocol. The EC signed on 29 April 1998, the US on 12 November 1998, and Japan on 28 April 1998. Those parties that have ratified the Protocol are Antigua and Barbuda, Bahamas, Bolivia, Cyprus, Ecuador, El Salvador, Fiji, Georgia, Guatemala, Jamaica, Maldives, Micronesia, Mongolia, Nicaragua, Niue, Palau, Panama, Paraguay, Trinidad and Tobago, Turkmenistan, Tuvalu, and Uzbekistan. 'Kyoto Protocol Status of Ratification (as at 13 January 2000)' at: <http://www.unfccc.de/resource/kpstats.pdf> as of 18 February 2000.

included in the Protocol.

The issues of the Protocol and the negotiation process are analysed below.

4.2.1 The Issues

This sub-chapter discusses the issues of the Kyoto Protocol in their finished and unfinished form. The Kyoto Protocol establishes quantified emission limitation and reduction commitments (QELRCs)[211] for all Annex I parties, specifying their assigned amounts, differentiation, the commitment period, the basket of gases, and the use of sinks to meet their commitments. The Protocol includes a list of policies and measures for Annex I parties to help achieve theirQELRCs. It establishes provisions for the so-called 'flexibility mechanisms', including joint fulfilment of commitments, joint implementation, emissions trading and the clean development mechanism. It failed to include provisions for voluntary commitments by non-Annex I countries. Finally, this section summarises the unresolved issues, and gives an overview of post-Kyoto progress to date.

4.2.1.1 Quantified Emission Limitation and Reduction Commitments

The main elements of the Kyoto Protocol's QELRCs are explained below.

Assigned Amounts: The issue of the assigned amounts, i.e., the numerical targets, was the most visible and comprehensible component of the Kyoto negotiations, and received the most public attention.[212] According to Article 3.1 of the Protocol, Annex I parties shall reduce their overall GHG emissions by at least 5 per cent below 1990 levels. The individual commitments of Annex I parties are listed in Annex B of the Protocol. The EC is thereby bound to a

211 Before COP3, the term QELRO (O = objective) was used as the general term for GHG targets and timetables.

212 See, for example, *The Japan Times*, 1 December 1997; and *International Herald Tribune*, 1 December 1997.

reduction target of 8 per cent, the US to 7 per cent and Japan to 6 per cent.[213]

Differentiation: Whereas the Convention had set a voluntary flat target for Annex I countries of stabilising emissions at 1990 levels by the year 2000, it had also mentioned the principle of common but differentiated responsibilities. The demand for differentiation can be deduced from this principle, but determining the level of responsibility of each country by calculus is a difficult and controversial undertaking. The Kyoto Protocol sets differentiated targets ranging from reductions of 8 per cent for the EC and other European states, including many Central and Eastern European states, to an increase of 10 per cent for Iceland.[214]

Commitment Period: The Berlin Mandate had called for commitments to be defined for time-frames such as 2005, 2010 and 2020. The Protocol adopted the concept of commitment periods. The first commitment period is set for 2008-2012, and negotiations on commitments for a second period are to start by 2005 (Article 3.9). Furthermore, each party 'shall, by 2005, have made demonstrable progress in achieving its commitments under this Protocol' (Article 3.2). The wording 'demonstrable progress', however, is not defined, which makes the provision rather weak.

Basket of Gases: Annex A of the Kyoto Protocol lists six gases: carbon dioxide, methane, nitrous oxide, hydrofluorocarbons, perfluorocarbons and sulphur hexafluoride. The baseline for the three main gases, i.e., carbon dioxide, methane and nitrous oxide, remains 1990. The Protocol allows any party to use a 1995 baseline for calculating the emissions of the other three gases.[215] The gases

---

213 Interestingly, the 5 per cent target matches the total decrease of GHG emissions from Annex I parties by 5 per cent since 1990, which was caused in large part by the collapse of the centrally-planned economies in the early 1990s and their subsequent decline in economic growth. During the same period, developing country emissions increased by 25 per cent. Missfeldt (1998), *op. cit.*, p. 128.

214 For background information and an overview of proposed indicators for differentiation by parties see FCCC/AGBM/1996/7.

215 The reason for this provision is that emissions of HFCs have increased in many Annex I countries between 1990 and 1995 as substitutes for CFCs, which are being phases out under the Montreal Protocol. It was necessary to allow a 1990 baseline for those parties (e.g., Norway), which had already started to phase out HFCs early. Furthermore, some countries have no data on emission levels of the latter three gases before 1995. Even

are taken together as a basket and compared according to their global warming potential (GWP), i.e., the radiative, or warming, impact of a molecule relative to carbon dioxide within a 100-year time horizon. This applies to the first commitment period, and the GWPs may be revised for subsequent commitment periods.[216]

Table 4 displays the vast differences of the gases in terms of their share of total emissions, on the one side, and their global warming potentials, on the other. The trace industrial gases account for a small proportion of total emissions. However, they are growing in importance because of their rising emissions and extremely long atmospheric lifetimes, and because they are used to replace CFCs.

Table 4: Greenhouse gases included in the Kyoto Protocol

Gas	**Sources**	**Emission trends since late 1980s**	**GWP-100**	**Emissions 1990, Annex I**
Carbon dioxide (CO_2)	Fossil fuel burning, cement	EU static, increases in other OECD countries, sharp decline in EITs	1	81.2 %
Methane (CH_4)	Rice, cattle, biomass burning and decay, fossil fuel production	Decline in most countries (big increase only in Canada, US and Norway)	21	13.7 %
Nitrous oxide (N_2O)	Fertilisers, fossil fuel burning, land conversion to agriculture	Varies, small increases in many countries, decline expected before 2000, decline in EITs	310	4.0 %
Hydrofluorocarbons (HFCs)	Industry, refrigerants	Fast-rising emissions due to substitution for CFCs	Around 1,300	0.54 %
Perfluorocarbons (PFCs)	Industry, aluminium, electronic and electrical industries, fire fighting, solvents	Static	Around 6,770	0.29 %
Sulphur hexa-fluoride (SF_6)	Electronic and electrical industries insulation	Increase in most countries, further rise expected	23,900	0.30 %

Source: adapted from Grubb et al. (1999), *op. cit.*, p. 73

though 1990 is a default base year, a lot of effort had already gone into establishing inventories for 1990. See also Grubb *et al.* (1999), *op. cit.*, pp. 71-72.

216 *Ibid.*, pp. 118-119.

Sinks: Sinks are carbon dioxide absorbing ecosystems, such as oceans, forests and soil. Large land areas reduce a country's net CO_2 emissions substantially. The issue of including sinks as 'negative emissions' had been proposed by the US as early as 1989.[217] Serious discussion on this matter was initiated by New Zealand during AGBM-8 in October 1997. The main difficulty concerning the issue is the high level of scientific, technological and methodological uncertainties. Furthermore, the long-term developments are still unclear; several studies point out that sinks can reach a saturation point, after which they begin to emit CO_2.[218]

The Kyoto Protocol allows for the inclusion of absorption by sinks resulting from post-1990 changes in forestry-related activities, including afforestation, reforestation and deforestation (Article 3.3), omitting for now agricultural soils and other carbon sinks. Furthermore, those Annex I parties for whom land-use change and forestry (LUCF) constituted a net source of greenhouse gas emissions in 1990 are allowed to include LUCF emissions when calculating their 1990 base year emissions, treating them almost like an additional source (Article 3.7).[219]

The inclusion of sinks distorts the actual scope of parties' commitments, considerably reducing the obligations for some parties. The US, for example, has effectively only to reduce its GHG emissions by 4%. According to OTT, the inclusion of sinks might one day be regarded as the biggest flaw of the Kyoto Protocol. It might undermine the Kyoto Protocol's verifiability and credibility, and thus erode its legitimacy and effectiveness.[220]

217 *Ibid.*, p. 76.

218 Kirk R. Smith, Joel Swisher and Dilip R. Ahuja (1993), 'Who pays (to solve the problem and how much)?', pp. 70-98, in: Peter Hayes and Kirk Smith (eds.), *The global greenhouse regime, Who pays?*, Tokyo, pp. 84-88. Grubb *et al.* (1999), *op. cit.*, pp. 24-25.

219 Yamin (1998), *op. cit.*, p 119.

220 Ott (1998), *op. cit.*. See also Henry D. Jacoby, Ronald G. Prinn and Richard Schmalensee (1998), 'Kyoto's Unfinished Business', pp. 54-66, in: *Foreign Affairs*, Vol. 77, No. 4, July/August 1998, p. 63.

4.2.1.2 Policies and Measures

The Berlin Mandate called for a strengthening of commitments as well as the elaboration of policies and measures (PAMs). They were to encompass any actions adopted by parties nationally, or internationally, to reduce emissions or enhance sinks. Article 2 of the Protocol includes a list ofPAMs, albeit in soft legal character, from which Annex I parties can choose in accordance with their national circumstances. The list suggests PAMs such as to improve energy efficiency, to protect and enhance sinks, to promote sustainable forms of agriculture, and to reduce market imperfections, fiscal incentives and subsidies that run counter to the objective of the Convention.[221]

The impetus behind the inclusion of PAMs is to guard the international competitiveness of those countries, or of specific industrial sectors within these countries, which have implemented PAMs unilaterally. In addition, somePAMs, such as the application of minimum standards of energy efficiency to traded products, or the taxation of aviation or marine bunker fuels, can only work effectively if implemented on a global scale.

4.2.1.3 Flexibility Mechanisms

The four flexibility mechanisms of the Kyoto Protocol comprise joint fulfilment of commitments, joint implementation, emissions trading and the clean development mechanism. They are outlined below.

Joint Fulfilment of Commitments (Article 4): Parties may fulfil their reduction commitments jointly, such as in the framework of a regional economic

221 The latter element interfaces with international trade, and thus has the potential for becoming the most heavily litigated provision in the Protocol under the binding dispute settlement procedures of the World Trade Organisation (WTO). See UNU, GEIC, IAS (1998), 'Global Climate Governance: A Report on the Inter-Linkages between the Kyoto Protocol and other Multilateral Regimes', Report – Part 1, Tokyo, pp. 9-16; and UNU, GEIC, IAS (1999), 'Global Climate Governance: Scenarios and Options on the Inter-Linkages between the Kyoto Protocol and other Multilateral Regimes', Report – Part 2, Tokyo, pp. 9-19. Furthermore, it reflects OPEC interests regarding the minimisation of economic impacts. Yamin (1998), *op. cit.*, p. 117.

integration organisation which is itself a party to the Kyoto Protocol.[222] The parties are merely required to notify the Secretariat and specify the respective obligations for each party to the agreement.

In case it fails to achieve its target, the organisation and each party to that agreement which has not achieved its target will be held responsible (Article 4.6). The organisation itself would thus be subject to sanctions, should individual countries fail to achieve their target, and should the parties establish a non-compliance procedure.[223]

The EU is the main benefactor of this provision, as well as its instigator. Its internal bubble agreement obliges the EU member countries to reduce emissions by 8 per cent jointly, with differentiated reduction – and even growth – targets for the individual member states that are negotiated internally.

Furthermore, this 'bubble' agreement will remain operative for the duration of a full commitment period, and any alteration in the composition of the organisation shall not affect the existing commitments. Thus, in the specific case of the EU, the accession of new member states would not change the EU's bubble arrangement for this period. As long as the EU achieves its overall reduction target, the EC and its member states are in compliance with the Protocol's provisions.

Joint Implementation (Article 6): As stated earlier, an agreement on joint implementation had not been possible at COP1 because industrialised and developing countries could not agree on whether the latter should be able to participate in the mechanism. Therefore, a pilot phase of 'activities implemented jointly' between any parties to the Convention was established to run until 2000, albeit without any crediting of achieved emission reductions.

Article 6 of the Protocol, although referring to 'projects' rather than explicitly mentioning 'joint implementation', enables emissions savings, or sink enhancement, arising from cross-border investments between Annex I parties, generating emission reduction units (ERUs). JI is to remain 'supplemental to domestic actions' (Article 6.1(d)). Article 3.10 and 3.11 provides for the

222 "Regional economic integration organization' means an organization constituted by sovereign States of a given region which has competence in respect of matters governed by this Convention or its protocols and has been duly authorized, in accordance with its internal procedures, to sign, ratify, accept, approve or accede to the instruments concerned' (Article 1.6, FCCC).

223 Ott (1998), *op. cit.*.

addition, or subtraction, of ERUs from assigned amounts, both for JI and emissions trading.

As JI, the way it was defined at COP3, takes place between parties that are subject to legally binding commitments, it does not entail any of the more political and technical complexities of JI between Annex I and non-Annex I countries described above. A party may entitle certain legal entities to participate under its responsibility in JI activities (Article 6.3). This paves the way for private sector participation, for which clear and verifiable guidelines still need to be established.[224]

Emissions Trading (Article 17): The emissions trading concept[225] had been announced by the US at COP2. The idea behind emissions trading is that parties can buy or sell emission allowances, thus making it easier for a party that faces higher costs or difficulty to achieve its assigned amount under the Protocol. Furthermore, emissions trading provides an incentive for companies and countries that find it relatively easy to cut emissions to go further than required. They would then profit by selling the excess to companies that find it harder or more costly to reduce emissions on their own. The approach thus achieves reductions efficiently and cheaply.[226]

Article 17 provides for the establishment of a trading scheme for Annex B parties for the purpose of fulfilling their commitments under Article 3. The Article begins by stating that related principles, modalities, rules and guidelines have yet to be defined by the Conference of the Parties, thus stressing the fact that emissions trading is still only a concept. Furthermore, the Article declares that trading shall be supplemental to domestic action, albeit not further defining its scope.

However, emissions trading introduces a major loophole: the so-called 'hot air' problem. Russia's CO_2 emissions are currently 30 per cent below 1990 levels and the Ukraine's are even lower, due to the sharp decline of their economies.[227]

224 Ott (1998), *op. cit.*.

225 The term 'emissions trading' is a little imprecise; it is not the emissions, but rather the *right* to emit, that is traded.

226 *International Herald Tribune*, 6-7 December 1997.

227 Russia accounts for 17.4 per cent of all CO2 emissions of industrialised countries in 1990. The Ukraine is the second largest emitter among the countries with economies in transition (CEITs), emitting around 30 per cent of the Russian share. The other successor states of the Soviet Union and the Central and Eastern European countries

In both cases, emission levels are expected to remain significantly below 1990 levels during the 2008-2012 commitment period, granting them a huge reduction surplus, given their obligations under the Protocol to merely stabilise their emissions at 1990 levels. This so-called 'hot air' could provide for cheap supply of emission credits for the industrialised countries, preventing some countries from undertaking any serious domestic action.[228]

Clean Development Mechanism (Article 12): The clean development mechanism (CDM) constitutes an amalgamation of two previously contradicting proposals.[229] It evolved from a Brazilian proposal for a clean development fund from June 1997. The proposal stipulates that Annex I parties should be subject to a financial penalty if they do not comply with their commitments under the protocol. The fine will be levied in proportion to the degree of non-compliance. It would be paid into a clean development fund that would be used to support emissions reduction or adaptation projects in developing countries.

In the final stages of the pre-Kyoto negotiations, it occurred to one or two 'key people' that the proposal could be interpreted to be something very similar to joint implementation.[230] If penalties for foregone emission reductions were levied at a rate that would finance projects in developing countries achieving the equivalent in emission reductions, then this would be practically identical to JI. The difference between the two would merely lie in their respective legal and institutional frameworks.

Subsequently, the idea of a penalty on governments was transformed into a mechanism for investment by companies.[231] The Protocol establishes the clean development mechanism in Article 12. It specifies that Annex I parties can earn certified emission reductions (CERs) from a CDM project in a developing country, and add these CERs to their assigned amounts under the Protocol.

There are two main features in the CDM that distinguish it from the earlier concept of JI. The first is the strength of multilateral control over the process in

---

emit far less in absolute terms, although energy efficiency is low in all these countries. Since the economic collapse of the former Soviet Union, Russian per-capita CO2 emissions have fallen by about 30 per cent. See Oberthür and Ot (1999), *op. cit.*, p. 21.

228 See Grubb *et al.* (1999), *op. cit.*, pp. 213-217.

229 It was called the 'Kyoto surprise'. Werksman (1998), *op. cit.*, p. 147. Grubb *et al.* (1999), *op. cit.*, p.101.

230 *Ibid.*, p.102.

231 *Ibid.*, pp. 101-103.

that it is to be supervised by an executive board and subject to the authority and guidance of the Conference of the Parties serving as the meeting of the Parties to this Protocol COP/MOP (Article 12.4)[232]. Secondly, a share of proceeds from project activities is to be used to cover its administrative expenses, as well as to assist particularly vulnerable developing countries in meeting the costs of adapting to climate change (Article 12.8).[233]

Furthermore, Article 12.10 states that CERs achieved between 2000 and 2008 may be counted towards a party's assigned amount in the first commitment period. However, the Article does not make a reference to CERs generated during the first commitment period, which, as GRUBB asserts, might be an omission in the drafting.[234]

The provisions of the CDM incorporate scope for global investment, and explicitly enshrine the role of the private sector (Article 12.9). The inclusion in the CDM of sinks remains disputed.[235]

4.2.1.4 Developing Country Participation

Despite the Berlin Mandate's explicit exclusion of the question of developing country participation from the protocol negotiations, the draft protocol included voluntary reduction targets and timetables for non-Annex I parties.[236] The draft article provided for non-Annex I parties that chose to commit themselves to a binding set of targets to select its historical base year, or period, and level of limitation. Upon acceptance by the Conference of the Parties, the party would then be bound by the same obligations as Annex I parties. Due to widespread developing country opposition this draft article was, in the end, omitted from the Kyoto Protocol.

Under the Protocol, developing country commitments are restricted to voluntary participation in the clean development mechanism and to the undertaking of general commitments according to Article 10 of the Protocol. This Article contains general obligations to formulate national programmes and

232 For details on COP/MOP see Yamin (1998), *op. cit.*, p. 124.

233 Ott (1998), *op. cit.*. Grubb *et al.* (1999), *op. cit.*, p. 133.

234 Grubb *et al.* (1999), *op. cit.*, pp. 133-135.

235 *Ibid.*, pp. 135-136.

236 FCCC/CP/1997/2.

develop political and scientific cooperation, along the lines of Article 4.1 of the Convention (see chapter 2.3.3).[237]

4.2.1.5 Unresolved Issues

In many cases, deferment of decisions to future meetings proved to be a constructive device for negotiating a protocol text that was balanced enough to reach consensus. This section outlines some of the many outstanding questions related to the Protocol, summarises the unresolved matters as contained in the Protocol and their timelines, and draws out the status of the Protocol details after COP5, which took place from 25 October – 5 November 1999, in Bonn.

Table 5: The Kyoto Protocol's outstanding decisions

Article	Details to be decided upon	Timeline for decision
2 (b)	Consider ways to facilitate cooperation with other parties to enhance the individual and combined effectiveness of the PAMs under Article 2 of the Protocol.	COP/MOP 1[238] or as soon as practicable thereafter
3.4	Decide upon modalities, rules and guidelines as to how, and which, additional human-induced activities related to changes in GHG emissions by sources and removals by sinks in the agricultural soils and the land-use change and forestry categories shall be added to, or subtracted from, the assigned amounts for Parties included in Annex I.	COP/MOP 1 or as soon as practicable thereafter
3.14	Consider what actions are necessary to minimise the adverse effects of climate change and/or the impacts of response measures on Parties referred to in paragraphs 4.8 and 4.9 of the Convention (i.e., developing countries that are particularly vulnerable to the adverse effects of climate change). Among these issues shall be the establishment of funding, insurance and transfer of technology.	COP/MOP 1
5.1	Decide upon guidelines for the creation of national systems for the estimation of anthropogenic emissions by sources and removals by sinks.	COP/MOP 1
5.2	Agree upon appropriate adjustments to the methodologies for estimating anthropogenic emissions by sources and removals by	COP/MOP 1

237 Ott (1998), *op. cit.*.

238 COP/MOP 1 means first session of the Conference of the Parties serving as the meeting of the Parties to this Protocol. See Article 13 of the Protocol.

	sinks that were not agreed upon at COP3.	
6.2	Elaborate guidelines for the implementation of joint implementation, including verification and reporting.	COP/MOP 1 or as soon as practicable thereafter
7.3	Determine the frequency of the necessary supplemental information required to be reported with the national inventories to ensure compliance with Article 3 of the Protocol	Some COP/MOP
7.4	Adopt and review periodically thereafter guidelines for the preparation of the information required under Article 7, taking into account guidelines for the preparation of national communications by Annex I Parties	COP/MOP 1
7.4	Decide upon modalities for the accounting of assigned amounts	Prior to the first commitment period
8.4	Adopt and review periodically thereafter guidelines for the review of implementation of this Protocol by expert review teams taking into account the relevant decisions of the COP	COP/MOP 1
9.2	Review commitments under the Protocol in the light of the best available scientific information and assessments on climate change and its impacts, as well as relevant technical, social and economic information.	COP/MOP 2 and at regular intervals thereafter
12.7	Elaborate modalities and procedures with the objective of ensuring transparency, efficiency and accountability through independent auditing and verification of project activities for the clean development mechanism.	COP/MOP 1
16	Consider the application to this Protocol of, and modify as appropriate, the multilateral consultative process referred to in Article 13 of the Convention.	As soon as practicable
17	Define the relevant principles, modalities, rules and guidelines, in particular for verification, reporting and accountability for emissions trading.	COP4 or future COP
18	Approve appropriate and effective procedures and mechanisms to determine and to address cases of non-compliance with the provisions of this Protocol, including through the development of an indicative list of consequences, taking into account the cause, type, degree and frequency of non-compliance.	COP/MOP 1

Source: developed from Barrett and Chambers (1998), *op. cit.*, pp. 88-90

The bulk of the 'unfinished business'[239] relates, above all, to the issue of sinks, the so-called Kyoto mechanisms (joint implementation, emissions trading and the clean development mechanism) and compliance.

239 See, for example, Ott (1998), *op. cit.*; and Jacoby, Prinn and Schmalensee (1998) *op. cit.*.

The term *sinks* (Article 3, KP) still lacks a distinct definition, especially in regard to the difference between 'natural' and 'anthropogenic' sinks, and to what constitutes a 'forest'. The language used is complex and incomprehensible. Furthermore, many technical questions, such as how to include the (uncertain) quantities of carbon stored below ground, will need to be addressed.[240] The 1998 Bonn meeting of the subsidiary bodies agreed on some of the technical clarifications of sinks,[241] which was adopted by COP4 in Buenos Aires, Argentina, in 1998.[242] The inclusion of additional categories of LUCF activities according to Article 3.4 is to be dealt with by the first session of the COP/MOP, and a Special Report on LUCF is to be prepared by the IPCC by 2000. Further decisions on this matter have been deferred until COP6.[243]

The three *Kyoto mechanisms* involve 'off-shore' activities for which rules, guidelines, verification and compliance measures have yet to be established. Many countries have demanded the limitation of such activities by setting caps, or imposing a transaction fee.[244]

The mechanisms allow for participation from the private sector, which requires the establishment of administrative structures. It is still not clear which institutions should be responsible for implementing the trading of credits, certification and monitoring of compliance. Article 6 (JI) allows parties to transfer or acquire emission reduction units (ERUs), but it is not specific about how exactly they are to be created. Parties also need to address how emission reductions are to be shared between donor and host countries.[245] These outstanding details are expected to be resolved by COP6.

Article 17 (emissions trading) fails to specify whether private sector entities should be allowed to trade – although it does so for the other two market

---

240 Yamin (1998), *op. cit.*, p. 119.

241 Grubb *et al.* (1999), *op. cit.*, p. 120.

242 Article 3.3, KP, means that 'the adjustment to a Party's assigned amount shall be equal to verifiable changes in carbon stocks during the period 2008 to 2012 resulting from direct human-induced activities of afforestation, reforestation and deforestation since 1 January 1990. Where the result of this calculation is a net emission, this value shall be subtracted from the Party's assigned amount'. FCCC/CP/1998/16/Add.1, Decision 9/CP.4.

243 Oberthür and Ott (1999), *op. cit.*, pp. 134-136.

244 Hermann Ott, Wuppertal Institute for Climate, Environment and Energy, 'Fachtagung: Globaler Klimaschutz am Scheideweg', Bonn, 11 October 1999.

245 Missfeldt (1998), *op. cit.*, p. 137.

mechanisms. It also fails to specify which gases should be tradable, and which private sector entities should be allowed to engage in emissions trading.[246] As emissions trading is explicitly assigned to the COP, it needs to be clarified how responsibilities between the COP and COP/MOP will be distributed on this issue.[247]

Article 12 (CDM) also leaves many questions unanswered. It is not yet clear whether the CDM will be applied to a portfolio of projects designed by developing countries to satisfy their specific needs, or whether it will be an accumulation of bilaterally agreed projects, constituting more of a governmental approach. Some industrialised countries, e.g., the US, will encourage participation of the private sector, which will, in turn, pursue its own interests.

With regard to the establishment of an executive board (Article 12.4), further specification is required concerning whether it would be a permanent body or one with periodic meetings, how many members it should comprise, and how they should be elected.

Another question is whether carbon sequestration projects should be allowed under the CDM. Connected to this is the question of how to measure the additional reduction compared to that which would occur in the absence of the certified project activity – especially for removal-by-sink activities.[248]

The Kyoto Protocol provides for the development of a non-*compliance* procedure by three means: Firstly, Article 16, KP, requests the COP/MOP to apply, and modify as appropriate, the multilateral consultative process established under the Convention in Article 13, FCCC, which has been approved by COP4.[249]

Secondly, Article 18, KP, calls for the approval of 'procedures and mechanisms to determine and to address cases of non-compliance with the provisions of this Protocol'. COP4 established a joint working group on compliance, which reported to COP5 that it plans to complete its work by COP6. Questions that remain open include the structure of the compliance body, whether it should have facilitative or enforcement functions, or both, and

246 *Ibid.*, p. 129.

247 *Ibid.*.

248 Estrada-Oyuela (1998 Advance Distribution Copy), 'First Approaches and Unanswered Questions', pp. 18-24, in: José Goldemberg (ed.), Issues & Options, The Clean Development Mechanism, UNDP, pp. 21-24.

249 Oberthür and Ott (1999), *op. cit.*, pp. 212-215.

whether an indicative list of consequences should be adopted, taking into account the cause, type, degree and frequency of non-compliance.[250]

Thirdly, Article 19, KP, incorporates the dispute settlement provisions of Article 14, FCCC, into the Protocol.[251] Although such provisions are commonly inserted into multilateral environmental treaties there has to date not been a single case in which parties to a dispute have resorted to a dispute settlement procedure.[252]

COP4, in November 1998, adopted the 'Buenos Aires Plan of Action',[253] defining the follow-up process. This two-year action plan establishes deadlines for finalising outstanding details of the Kyoto Protocol to make it fully operational for entry into force after COP6.[254] It includes six specific areas:

- A financial mechanism;
- Development and transfer of technologies;
- Addressing the specific needs of developing countries (referred to in Article 4.8 and 4.9 of the Convention);
- AIJ under the pilot phase;
- The Kyoto mechanisms; and
- Preparations for COP/MOP1.[255]

Also, Argentina proclaimed that it intended to take on voluntary commitments at COP4.[256] It announced its adoption of a voluntary target to reduce its GHG emissions by 2-10 per cent in the 2008-2012 period at COP5 in 1999, without abandoning its status as a non-Annex I party.[257]

---

250 *Earth Negotiations Bulletin*, 8 November 1999.

251 This dispute settlement is modelled after Article 33 of the United Nations Charter. Oberthür and Ott (1999), *op. cit.*, p. 222.

252 Grubb *et al.* (1999), *op. cit.*, pp. 148-149. Oberthür and Ott (1999), *op. cit.*, pp. 222-223.

253 FCCC/CP/1998/16/Add.1, Decision 1/CP.4

254 Bharat H. Desai (1999), 'Institutionalizing the Kyoto Climate Accord', pp. 159-163, in: *Environmental Policy and Law*, 29/4 (1999), p. 161.

255 For further details see Grubb *et al.* (1999), *op. cit.*, pp. 249-253.

256 This move was followed by the US signing the Kyoto Protocol. There are rumours, however, that Argentina's announcement is connected to US payment of the conference costs, which are usually paid by the host country. Yet, it opened up the prospect of a new negotiation process toward binding commitments by developing countries.

257 *Earth Negotiations Bulletin*, 16 November 1998. Mariana Conte Grand (1999), 'Argentina's Voluntary Commitment Proposal at COP5: What is it About?', Weathervane, At the Negotiating Table, December 1999, at:<http://www.weathervane.rff.org/negtable/Argentina's_Proposal.html>.

COP5 achieved its work plan ahead of schedule and thus parted in an atmosphere of optimism. It adopted a set of 22 decisions, which focussed primarily on procedural issues and prepared for decisions at COP6, scheduled for November 2000 in The Hague, The Netherlands, on the Kyoto mechanisms and compliance.

4.2.2 The Process

This sub-chapter sketches the key shifts in the various positions and the negotiation dynamics during COP3 towards adoption of the Protocol. It illustrates how the parties bargained and gave ground in order to reach agreement, and in what manner agreement was ultimately reached. The issues are (1) QELRCs encompassing assigned amounts, differentiation, the commitment period, the basket of gases and sinks, (2) policies and measures, (3) flexibility mechanisms including joint fulfilment of commitments, joint implementation, emissions trading and the clean development mechanism, and (4) developing country participation. Unless specified otherwise, the information of this section is derived from an audio recording of the sessions.[258]

The negotiations were conducted in plenary sessions as well as in negotiating and informal groups, which reported back to the plenary at regular intervals. The COP established the Committee of the Whole (COW) – an ad hoc working group – at its first meeting on 1 December. It was chaired by the former AGBM Chair Raul Estrada-Oyuela, the Argentinean Ambassador to China.

The COW's task was to consider the COP agenda item on the adoption of a protocol; it was thus responsible for the final round of negotiations. During the final COW session, from the evening of 10 December until its adjournment in the late morning of 11 December 1997, the Chair's final draft[259] was reviewed and adopted article by article. The COP met shortly thereafter to adopt the Kyoto Protocol by consensus.

---

[258] The audio recording is kept in the UNFCCC Secretariat library in Bonn.

[259] Document FCCC/CP/1997/CRP.6. See also *Earth Negotiations Bulletin*, 13 December 1997.

4.2.2.1 Quantified Emission Limitation and Reduction Commitments

Assigned Amounts: The target figures, i.e., assigned amounts, became highly politicised, and were the most visible element of the Protocol. Movement on the issue began on 3 December when the US indicated that it might introduce a 'small symbolic percentage cut' of GHGs, which went beyond its official proposal of stabilising emissions at 1990 levels.[260] Chair Estrada took the initiative to compile the target figures, due to the explosive nature of the issue and since the parties made the setting of their targets dependent upon the targets of others.

On 6 December, he proposed that the EU should reduce its emissions by 10 per cent, the US by 5 per cent and Japan by 2.5 per cent. The EU was not supportive of this proposal and maintained that the EU, US and Japan should have the same target.[261] The issue had to be suspended until the ministerial segment starting on 8 December, at a point when the EU required a ministerial decision for further negotiations on the targets issue.[262]

On 8 December, US Vice President Al Gore made a 12-hour trip to the Kyoto Conference where he instructed his delegation 'to show increased negotiating flexibility if a comprehensive plan can be put into place'.[263] This message may have swayed US negotiators' willingness to accept higher targets.

On 9 December, Estrada merged the proposed targets for the EU, US and Japan slightly to 8 per cent, 5 per cent and 4.5 per cent, respectively.[264] Finally, on 11 December, he circulated the 8-7-6 per cent figures, contained in Annex B, which was ultimately adopted.

With 8 per cent of 1990 levels, the EU retains it leadership commitment, albeit on a very narrow margin. The US consented to a 7 per cent reduction

260 The issue of symbolic reductions had first arisen during informal bilateral talks between the US and Japan earlier in the week. *The Japan Times*, 4 December 1997.

261 *Süddeutsche Zeitung*, 8 December 1997. *The Daily Yomiuri*, 8 December 1997.

262 *Earth Negotiations Bulletin*, 3 December 1997.

263 Remarks by Vice President Al Gore at the COP3 high-level segment, 8 December 1997, on file with author. This sentence was not in an advance copy of his speech, and was interpreted as ambiguous by many commentators. For comments on Gore's speech made by the press see, for example, *International Herald Tribune*, 9 December 1997; *The Japan Times*, 9 December 1997; *Financial Times*, 9 December 1997; and *Süddeutsche Zeitung*, 9 December 1997.

264 *The Daily Yomiuri*, 10 December 1997.

target, which is a substantial commitment given its increase in emissions since 1990 of about 23 per cent. Japan agreed to a reduction target of 6 per cent of 1990 levels, which might be the most difficult one to achieve. It is widely believed that Japan might not have consented to this target if it had not hosted COP3.[265]

Differentiation: The US announced on 1 December that it was willing to accept a differentiated approach for all countries.[266] Japan followed the announcement by declaring on 2 December that the EU, US and itself should be counted as three separate categories of Annex I countries, in order to reflect an equitable burden on the basis of past achievements. This approach would take into account the EU's clear advantage from its individualistic composition, and the US's much higher energy consumption on a per-capita ratio compared to any other country. Recognising Japan's past achievements and its resulting extreme energy efficiency, it should subsequently be given a lower reduction target than the EU or the US.[267]

The EU remained firm on its position that Annex I parties should have the same target, until 7 December. It then indicated that it might compromise on differentiation for other Annex I countries if at least the EU, US and Japan would pursue the same initial target. Otherwise, the EU would be faced with a competitive disadvantage compared to the US and Japan. The matter was ultimately settled by introducing differentiation among the three, but at a margin of 1 percentage point differences which in effect comes very close to a flat rate as demanded by the EU.

Commitment Period: The US, with the support of Japan, negotiated for a multi-year target of five years; the first was to span from 2008-2012. The EU, in contrast, proposed 2010 as a single-year target.

The US argued that such a multi-year target, or 'budget period', would provide flexibility to the parties in the event of unforeseen fluctuations in economic performance, or extreme weather conditions, prior to a specific target year. The period approach would thus allow for an overshooting of the target in a single

265 Personal communication with Joanna Depledge, 26 January 2000.

266 *ECO*, 2 December 1997. *The Japan Times*, 2 December 1997. The US had advocated stabilisation of emissions at 1990 by 2000 for all Annex I countries before COP3.

267 'Press Briefing by the Government of Japan, COP3, 2 December 1997', at: <http://www.mofa.go.jp/policy/global/environment/warm/cop3/session.html> as of 9 July 1999.

year.[268] After the EU had accepted the period approach, it attempted to strengthen the commitments by pushing the proposed period of 2008-2012 down to 2003-2007.

On 9 December, Chair Estrada suggested a compromising 2006-2010 period. However, by then EU support for an earlier period had diminished, as it would reduce numerical commitments, making the final outcome of COP3 appear even less ambitious.[269]

The US managed to sell its concept but the term 'budget period' was replaced with 'commitment period'. Many delegates had expressed concern as the word 'budget' could allude to the concept of emissions borrowing.The EU and most developing countries had opposed this concept because it presented a potential loophole.[270] As a further consolation to the EU, Estrada's soft provision that 'each Party ... shall, by 2005, have made demonstrable progress in achieving its commitments under this Protocol' (Article 3.2) was kept in the text.

Basket of Gases: While the EU and Japan favoured the inclusion of only three greenhouse gases (CO_2, CH_4 and N_2O) in the so-called basket of gases, the US argued for a comprehensive approach incorporating six gases (the three above plus HFCs, PFCs and SF_6), arguing that this was the more environmentally sound option. One reason why the EU, in particular, was against the inclusion of the three additional gases was its fear that they might provide loopholes in the agreement, as the methodologies for inventorying these gases are not yet fully developed.[271]

In order to overcome this divide, Chair Estrada suggested a two-basket approach, by which a three-gas basket would be accepted at this stage, and the remaining three gases would follow at a later date. The US expressed concern that it would require another negotiating process, which could be avoided if all six gases were included at COP3.[272]

The negotiations were driven primarily by economic and political arguments

268 Ott (1998) *op. cit.*.

269 Grubb *et al.* (1999), *op. cit.*, p. 70. *The Japan Times*, 10 December 1997.

270 Barrett and Chambers (1998), *op. cit.*, p. 57.

271 However, this was not the only reason. See Grubb *et al.* (1999), *op. cit.*, pp. 72-76, 119, for a more detailed analysis of the underlying incentives of the EU, US and Japan in choosing their numbers of GHGs.

272 'US Delegation to COP3, Press Briefing, Kyoto, Japan, 3 December 1997', at: <http://www.state.gov/www/global/oes> as of 30 October 1998.

for and against including these gases. CH_4 and N_2O, for instance, were widely endorsed because emissions of these gases are currently stagnating, or even declining, in most countries (see *Table 4*, p. 49). It ultimately allowed some countries, such as the US, to consent to stronger emission targets.[273]

On 10 December, the US succeeded in convincing the parties to place six gases in the basket, contained in Annex A to the Protocol. As a concession to Japan and the EU, the Protocol provides that any party may use a 1995 baseline for calculating the emissions of SF_6, HFCs and PFCs.[274]

Sinks: The sinks issue was another controversial item. It was supported by countries with large land areas, basically the 'New World' countries and Scandinavia. Opponents such as the EU – despite some internal frictions –[275] as well as the G77 & China argued that its inclusion would provide another means of weakening the commitments of some parties. These parties would receive unearned credit for naturally occurring absorption.

Moreover, including sinks could induce these countries to undertake environmentally unfavourable practices such as replacing mature, old-growth forests with fast-growing monoculture plantations to enhance their removal-by-sinks record. There was also the so-called 'Amazonian picket fence' problem, meaning that parties could simply erect a fence around natural forests and take credit for it. A great deal of discussion revolved around the definition of such 'anthropogenic sinks'.[276]

Moreover, the fact that the uptake of CO_2 by sinks is still very difficult to monitor accurately was also held against its supporters.[277] The issue required an early solution so that parties could calculate their individual targets; it had a significant impact on the target numbers, and it was one of the reasons for the US to consent to higher reduction targets.[278]

On 9 December, the COW decided to allow only increases in GHG removal by sinks due to afforestation, reforestation and deforestation activities since

273 Grubb *et al.* (1999), *op. cit.*, p. 119.

274 Yamin (1998), *op. cit.*, p. 118.

275 Grubb *et al.* (1999), *op. cit.*, p. 80.

276 Yamin (1998), *op. cit.*, p. 118.

277 See also Grubb *et al.* (1999), *op. cit.*, pp. 76-79. Parties' views on sinks, submitted only a few weeks before COP3, can be found in documents FCCC/AGBM/1997/INF.2 and FCCC/AGBM/1997/MISC.4 and Add.1.

278 *Earth Negotiations Bulletin*, 5 December 1997.

1990 to be added to a party's amount of GHG reductions in the first commitment period.[279] Furthermore, they agreed to delete the whole category of land use change and forestry (LUCF) from the list of sectors and sources contained in Annex A of the Protocol. This marked a partial victory for the EU. Furthermore, the parties agreed to initiate a process to eventually cover all categories of sinks, and thus provide a comprehensive scheme for the second commitment period.[280]

4.2.2.2 Policies and Measures

The EU argued strongly for a coordinated approach to PAMs, including a long and detailed set of mandatory and voluntary policy options. This had been a key element of its negotiating strategy. The US, in turn, argued that it should be up to each party to choose its own PAMs in tandem with national circumstances.[281] Japan suggested allowing the parties to voluntarily choose PAMs listed in the protocol, and elaborate internationally agreed indicators to assess whether a party's target was being met. Thus, the US and Japan supported a far more *laissez-faire* position, without any constraints on policy choice.[282]

The parties also debated whether PAMs should apply to non-Annex I parties, and concluded that they should not.[283] The Article was adopted during the final COW session with almost no further debate. It represented more of a victory to the US and Japan than the EU, because the wording of Article 2.1(a), requesting parties to adopt general measures 'such as', replaced the EU wording of 'in particular' of the previous draft, thus presenting only a menu of policy options to choose from.

4.2.2.3 Flexibility Mechanisms

Joint Fulfilment of Commitments / The EU Bubble: Many parties expressed criticism of the EU bubble. The US criticised the lack of transparency and accountability inherent in the concept. It requested clarification on questions

279 *Earth Negotiations Bulletin*, 13 December 1997.

280 Yamin (1998), *op. cit.*, p. 118.

281 Grubb *et al.* (1999), *op. cit.*, p. 65. Yamin (1998), *op. cit.*, p. 116.

282 'Press Briefing by the Government of Japan to COP3, 3 December 1997', *op. cit.*.

283 *Earth Negotiations Bulletin*, 13 December 1997.

concerning how EU enlargement would affect the bubble, and on the specific legal relationship between the European Commission and the EU member countries on this issue. Moreover, the US interpreted the bubble to be a trading mechanism – an interpretation the EU did not support – and criticised the EU for denying the right to trade emission credits to other countries outside the EU.[284]

Japan also pointed to the unfairness regarding the EU bubble vis-à-vis non-EU countries in that it opposed the principle of differentiation for countries outside the EU. Furthermore, Japan demanded that the legal responsibilities should be clarified, and that the question of equity between EU members and non-members should be discussed. Japan saw in the bubble provisions for 'tremendous' differentiation – which, in principle, it welcomed. Japan argued that it would be easier for countries within the EU bubble to achieve their obligations, and the burden of each EU country could be modified relatively easily. Thus it proposed that the EU should assume a greater degree of responsibility, i.e., abide by a higher target than other Annex I parties. This would be a way of maintaining equity in burden-sharing.[285]

EU Environment Commissioner Bjerregaard conceived this line of argument as 'completely unacceptable, a nonstarter, a misunderstanding of what the EU is'.[286] In part, the criticism the EU received for its bubble concept was due to its own shortcomings. It did not present the legal aspects of its concept to the other parties before AGBM-8, thus leaving room for questions.

Moreover, the EU never satisfactorily explained why its approach was equitable. If one looks at the absolute and per-capita figures of the member countries' GHG emissions, it becomes evident that those countries that are allowed to increase their emissions by a substantial margin are those, which produce low figures.[287] In the end, the bubble concept was adopted without further discussion during the last COW session. It was accepted under the compromise that the COP would not have to approve the EU burden sharing

284 'US Delegation to COP3, Press Briefing, Kyoto, Japan, 4 December 1997', *op. cit.*.

285 'Press Briefing by the Government of Japan at COP3, 1, 3, 4 and 7 December 1997', *op. cit.*.

286 *The Daily Yomiuri*, 5 December 1997.

287 For an examination of emission characteristics in the EU member states see Collier (1997), *op. cit.*, pp. 44-47. While a 25 per cent reduction of emissions in Germany accounts for a reduction of 253.25 million tonnes of CO_2, Portugal's 40 per cent increase would merely add 17 million tonnes to the EU's total CO_2 emissions. Ringius (1997), *op. cit.*, p. 46.

arrangement, and that the EU could not change that arrangement once it had been put in place.[288]

Joint Implementation: The EU favoured limiting JI to Annex I parties, and suggested to await the results of the pilot phase before extending JI to include developing countries.[289] The G77 & China had requested the deletion of this draft article as they feared that Annex I parties would achieve their emission targets without taking adequate domestic action, thus escaping their primary responsibility. Furthermore, they feared that Annex I parties would be able to use up the cheap options of emission reductions in developing countries, and that they would be left with the more expensive reduction options once they came to adopt emission reduction commitments themselves.[290]

After industrialised countries had agreed on their reduction obligations, and developing countries were excluded from binding commitments, JI with crediting among Annex I parties was adopted during the final COW session without further discussion. Due to EU insistence, it was to be 'supplemental to domestic actions' (Article 6.1(d)); however, this supplement was not further defined.

Emissions Trading: The debate over emissions trading was polarised between developed and developing countries. The EU supported emissions trading because of the leeway it provides for higher overall reduction targets, and the US and Japan for its cost-effective nature. The US had made its inclusion a key prerequisite for its adoption of a protocol. The developing countries, on the other hand, held that it was a concept they did not fully understand, and regarded it as counter to the goal of the Berlin Mandate.

Furthermore, the developing countries suspected emissions trading to be a potentially cheap means for countries such as the US to 'buy' themselves out of their obligations. Because of this, there were calls from within the EU to place an upper ceiling or 'cap' on the percentage of a country's obligation that can be achieved through emissions trading.

Japan's concern was that the US could use its enormous political leverage over Russia to monopolise its emissions surplus. It thus demanded the scheme to be

288 Personal communication with Joanna Depledge, 26 January 2000.

289 *Frankfurter Allgemeine Zeitung*, 29 November 1997.

290 Missfeldt (1998), *op. cit.*, p. 134. For a more detailed account of North-South debates on the issue see Grubb *et al.* (1999), *op. cit.*, pp. 97-100.

transparent and competitive. There was also some disagreement about the starting date of emissions trading. While the US believed that it should start immediately and would not require further action by the Conference of the Parties, the EU demanded that clear rules and guidelines be established before initiating such a mechanism.[291]

During the last COW session in night of 11 December, after a fierce debate that lasted for about two hours, a compromise solution was settled upon. At around 4 a.m., Chair Estrada proposed to omit the sub-article[292] and replace it by a short interim article 16bis – which had been submitted by the UK on behalf of the EU – and which later became Article 17.

After many parties, including developing countries, had expressed their support for the UK amendment, Chair Estrada decided that its incorporation into article 16bis was the only consensus possible. He quickly wielded his gavel and adopted the Article. According to GRUBB, he ignored India, and possibly other parties, who had raised their placards, and thus overrode any possible objections. None of these countries, however, challenged Estrada's authority on this point.

Finally, despite earlier fierce opposition by the developing countries, the concept of emissions trading was incorporated into the final text.[293] This was, above all, a victory for the US. The EU managed to insert its provision that the rules and guidelines for verification, reporting and accountability should be developed first and be supplemental to domestic action to reduce emissions. This provided the EU with a platform to pursue its demand for an emissions ceiling.

Clean Development Mechanism: The EU remained opposed to the CDM, seeing it as both a potential loophole and a competitor to the Global Environment Facility (GEF). It was especially suspicious of the draft article's clause on early crediting between 2000 and 2008, and thus 'banking' of certified emission reductions[294] – unlike the US and Japan which were in favour of it.

Within 48 hours of negotiations behind closed doors by an informal contact group, just before the end of COP3, the proposal evolved from an attempt to create a compliance mechanism to a mechanism facilitating project-based joint implementation between developed and developing countries. When it came up

---

291 Ott (1998), *op. cit.*. Yamin (1998), *op. cit.*, p. 122.

292 Draft article 3.10, FCCC/CP/1997/CRP.6.

293 Grubb *et al.* (1999), *op. cit.*, pp. 95-96. *Financial Times*, 11 December 1997.

294 Draft article [illegible] FCCC/CP/1997/CRP.6.

for final agreement during the last COW session at around 7 a.m., most parties had not seen the final text.[295]

When the Chair called out draft article 13.10 for adoption, the Philippines, Marshall Islands and Uganda took the floor to object, while Russia, Costa Rica, Hungary, the US and Canada spoke out to retain it. The EU failed to raise its placard in time to object, due to internal organisational delays. The Chair thus ruled that there was not enough opposition to delete an instrument that some delegations considered useful. Several EU member countries took the floor to object, but Chair Estrada refused to concede. His consolation to the EU was to refer this point to the subsidiary parties and suggest that an analysis of the impacts of early crediting be taken up at COP4.[296]

4.2.2.4 Developing Country Participation

The US had strongly insisted on enshrining provisions for developing country participation in the protocol.[297] It became one of the most heatedly debated issues of COP3. On 30 November, prior to COP3 (resumed AGBM-8), Chair Estrada expressed his strong opposition to new commitments for developing countries. He emphasised that developing countries had made clear progress in implementing measures to reduce emissions growth, and urged industrialised countries that had not fulfilled their own commitments under the Convention to bear this in mind.[298]

---

295 Joanna Depledge (no date), 'From Kyoto to Buenos Aires: Prospects for COP4', pp. 1-6, in: W. Bradnee Chambers (ed.), *Global Climate Governance, Inter-linkages between the Kyoto Protocol and other multilateral regimes*, Tokyo.

296 Article 5(e), FCCC/CP1997/7/Add.1, Decision 1/CP.3. See also Grubb *et al.* (1999), *op. cit.*, p. 135.

297 The US stated at various instances that it favoured differing targets for developing countries, including emissions growth targets for some. *Earth Negotiations Bulletin*, 13 December 1997.

298 These findings are compiled in document FCCC/AGBM/1997/CRP.5. The report lists the various activities undertaken in countries such as China, India, Indonesia, Thailand and the Philippines. It observes that climate protection is not the sole reason for these policies and measures, but that concerns over public health, air pollution, job creation and economic effectiveness play an equally important role. *ECO*, 1 December 1997. *Earth Negotiations Bulletin*, 1 December 1997. See also Desai (1999), *op. cit.*, p. 161, p. 163 n. 24.

Various proposals on the issue were subsequently circulated.[299] The most prominent one was submitted to the COP by New Zealand on 5 December, which was supported by the EU, US and Japan. The proposal sought to start a process towards commitments from non-Annex I countries to limit emission growth – not reduce emissions –, starting in 2014. It would apply to all but the least developed countries.[300]

The EU position was that, in line with the IPCC findings indicating that a significant reduction in emissions would require efforts by both developed and developing countries, a review process to establish further commitments for all countries should be launched, but only after developed countries had adopted legally binding commitments. Japan pointed to the need for further participation in future and proposed a post-Kyoto process to this effect.[301]

Although the New Zealand proposal was a mild one relative to the US position on the issue, it did end negotiations for that day after thirty developing countries had taken the floor to protest against it,[302] conveying to the industrialised countries their principle: 'until you deliver, we don't discuss'.[303]

During the last COW session, the issue received some further heated attention. 14 countries requested its deletion,[304] whilst 13 supported its inclusion, with or without amendments to the text,[305] of which one spoke on behalf of AOSIS (35 countries) and one on behalf of the EU (15 countries).[306] The main opponents were China, India and Brazil, as well as the OPEC countries, which had been advised by the fossil fuel industry that voluntary commitments from developing countries would significantly improve the US Senate's likelihood of ratifying the protocol.[307] Despite the large number of proponents of draft article 10 and its soft language, Estrada decided at around 6 a.m. that it was too late for further

299 Yamin (1998), *op. cit.*, p. 123.

300 *The Daily Yomiuri*, 7 December 1997.

301 *Earth Negotiations Bulletin*, 13 December 1997.

302 Ott (1998), *op. cit.*.

303 *Earth Negotiations Bulletin*, 13 December 1997.

304 They are Saudi Arabia, Brazil, Kuwait, Iran, Venezuela, India, China, Egypt, Uganda, United Arab Emirates, Syria, Morocco, Gambia and Bangladesh.

305 Draft article 9, FCCC/CP/1997/CRP.6. Argentina, the US, Korea and Mexico proposed amendments to the text.

306 They are Samoa (AOSIS), Argentina, Hungary, the US, Grenada, Korea, Japan, Russia, Philippines, Israel, Mexico, (Micronesia) and the UK (EU).

307 Oberthür and Ott (1999), *op. cit.*, p. 230.

debate and the consideration of new proposals. As consensus seemed impossible, he decided to simply delete the article.

4.2.3 The Outcome

The Kyoto Protocol was ultimately a political deal. The final outcome of the negotiations reflects a patchwork of different initial positions and proposals. GRUBB characterises the Protocol as an agreement 'to establish a global emissions market to counter the global environmental consequences of global economic growth'.[308] Moreover, KAWASHIMA contends that, during the FCCC process, the EU has taken on the role of a leader, Japan that of a follower, and the US that of a laggard.[309]

This section examines the outcome of COP3 in respect to the priorities and strategies of these three actors. It attempts to shed some light on the motivations behind their changes in position, insofar as they have been disclosed by insiders and reflected upon in the literature. Importantly, this section attempts to explain the outcome of those issues that are not attributable to the priorities, or strategies, of the three actors, but rather to certain dynamics during the final hours of the Conference.

308 Grubb *et al.* (1999), *op. cit.*, p. 137.

309 Kawashima (1997), *op. cit.*, p. 96.

Table 6: The outcome of the negotiations

EU position	US position	Japanese position	Outcome
Targets and timetables			
-15% by 2010; -7.5% by 2005	Stabilisation by 2008-2012	-5% by 2008-2012	At least -5% by 2008-2012 (Article 3.1)
Differentiation			
Burden sharing within the EU; no differentiation outside the EU	Differentiation for all (since 1 December 1997)	Differentiation depending on a country's emissions per capita + GDP and population growth	Differentiation for all Annex I (AI) countries (Annex B)
Basket of greenhouse gases			
3 GHGs	6 GHGs	3 GHGs	6 GHGs (with the latter 3 of 1995 levels) (Annex A)
Sinks			
To be decided by COP/MOP	In favour	In favour	Included but limited for now to afforestation, reforestation and deforestation since 1990 (Article 3.3 and 3.7)
Polices and measures			
COP should set mandatory PAMs	Should be up to each party individually	Should be up to each party individually	Menu of PAMs (Article 2)
EU bubble			
In favour	With reservations	With reservations	Included (Article 4)
Joint implementation			
Only for AI parties until the end of the pilot phase	Strongly in favour	Only under certain conditions	Included (Article 6), for AI parties only
Emissions trading			
Opposed	Strongly in favour	Only under certain conditions	Included (Article 17), with references in Articles 3.10 and 3.11); rules to be negotiated
Clean development mechanism			
Opposed	In favour	In favour	Included (Article 12), under supervision of an executive body and COP/MOP
Developing country participation			
No commitments for developing countries (DCs)	Meaningful participation by key DCs	No commitments for DCs	Decision postponed to future COPs

Source: developed from Gupta and Grijp (forthcoming), *op. cit.*.

4.2.3.1 Strategic Determinants

Given its ambitious target, the EU pushed strongly for higher percentage reductions among Annex I parties, primarily the US and Japan. Without these efforts, the final set of targets would have undoubtedly been weaker. Although the targets are differentiated, the margins between the EU, US and Japanese assigned amounts are so narrow that they will not affect the EU's, or Japan's, economic competitiveness. The US has to reduce its emissions by around 30 per cent in real terms, because of its substantial increases in emissions since 1990. This outcome can be seen as a sign of Al Gore's appeal to the US delegation for increased flexibility. However, this also puts into question whether the US will achieve its target within the first commitment period.

These reduction targets subsequently served as the engine of compromise on other outstanding issues, notably sinks and the flexibility mechanisms. The US was inclined towards maximum flexibility in implementation measures. It succeeded in including not only some unwieldy language on sinks and mechanisms for international transfer among Annex I parties, but also a clean development mechanism enabling participation of non-Annex I parties.

The EU managed to integrate the EU bubble into the agreement. The EU's provisions to restrict the level of the 'off-shore' activities were inserted to address potential loopholes in the agreement. As a result, these mechanisms are subject to supplementarity, supervision and verification. Furthermore, the US managed to include in the Protocol six GHGs instead of only three. The US also succeeded in including its five-year commitment period spanning from 2008-2012. This five-year window for compliance is expected to help lower costs and average the effects of fluctuations in business cycles and energy demands. Moreover, it allows one full decade for transition to more energy-efficient technologies.[310] Although the EU obtained its desired list of policies and measures, these were substantially watered down.

The EU position on high reduction targets and the EU bubble, which was regarded by many delegates as hypocrisy, was part of its negotiating strategy.

---

310 Richard A. Horsch and Joseph D. Richards (1998), 'Does Kyoto Protocol Fall Short of the Mark?', New York Law Journal, at: <http://www.ljx.com/practice/internat/0427kyoto.html> as of 27 April 1998.

The EU is said to have developed its position in advance of the others and negotiated towards it.[311] The pressure by others on the EU bubble was also a tactic; its intention was to help extract concessions from the Europeans on other issues, such as differentiation for all countries and emissions trading.

US delegates have disclosed that the general reason behind America's readiness to accept a compromise in the end was that it did not want to be held responsible for the failure of Kyoto. Furthermore, the administration wished to present an environmental profile that conformed to its democratic self-image. It also felt the need to respond to the climate change science, which is primarily being developed in the US. Finally, opinion polls indicated that the American public was less indifferent to the climate change issue than previously recognised.[312]

Japan made constant efforts to find a middle road between the two above positions.[313] At no point did Japan adopt a position opposing both the EU and US. It consented to higher targets than initially advocated, but this was because its ultimate priority lay in reaching consensus for a protocol, regardless of the exact details.

It is important to note that, despite much academic debate about the decline of the nation-state in the era of globalisation, the Kyoto Protocol was ultimately fought out by governments. Except for in the COW sessions, the various NGO constituencies were not present when delegates met in closed meetings and hotel rooms to strike the delicate balance of consensus that was required for the adoption of a protocol.[314]

4.2.3.2 Dynamics

Given the specific positions and strategies of the three key players, there were two surprise outcomes to the Protocol: the inclusion of a clean development mechanism and the exclusion of voluntary commitments from developing countries. These issues could not be settled earlier, and were only decided

311 Gupta and Grijp (forthcoming), *op. cit.*.

312 *Ibid.*

313 Barrett and Chambers (1998), *op. cit.*, p. 54.

314 For an account on participation of NGOs in the UNFCCC process see Glen Paoletto and Heike Schroeder (1997), 'Enhancing Participation of NGOs in the UNFCCC Process', GEIC Paper Series, October 1997.

during the last hours of the final COW session. As a result, additional factors impacted on the resolution of these issues.

The *clean development mechanism* has been called the 'Kyoto surprise'[315] because it evolved in closed negotiations at the very end of the conference and was unwrapped during the final COW session. In hindsight, it presents a clever and equitable means of involving developing countries in the Kyoto Protocol. Yet, it was adopted despite the fact that most parties had not seen the final text, and thus had not had a chance to contemplate its details. Given that it was already 7 a.m. and time was pressing, Estrada allowed little time for debate on this issue.

The most controversial aspect to the CDM was early crediting starting in 2000 (draft article 13.10). It had been addressed by 8 parties when Estrada suddenly wielded his gavel to adopt the draft article, which became Article 12.10 of the Protocol. The EU had taken longer to agree internally that it was opposed to draft article 13.10 than the COW, and thus missed its chance to inform the chair of its position.[316] The Philippines, the Marshall Islands and Uganda had supported its deletion, while Russia, Costa Rica, Hungary, the US and Canada marshalled a stronger bloc to retain it. Had the EU, with the weight of 15 parties, made its statement earlier, it is possible that the Chair would have ruled differently. Because of Estrada's self-imposed ruling that he could not reopen an article once a decision was taken, it was clearly not possible to reconsider this item despite the protest of various EU member states.

This internal delay was in part caused by the delegation structures under which the EU operated. Since the EU is not a full party to the Convention, the articulation of its negotiating positions and proposals remains in the hands of its member states. The country holding the 6-month EU Presidency plays an important role in enforcing and coordinating its negotiating strategies. The so-called 'troika', made up of the country presently holding the EU Presidency, together with the one immediately before and the one immediately after, takes the lead in this regard, and one designated member state, usually the one holding the Presidency, speaks on behalf of the EU. However, the larger EU member countries are reported to have disrespected this structure.[317]

The lack of a natural hierarchy causes inflexibility and slow responsiveness,

---

315 See footnote 224.

316 Oberthür and Ott (1999), *op. cit.*, p. 90.

317 See, for example, Oberthür and Ott (1999), *op. cit.*, p. 86.

as can be seen above. Delegates have described the EU as overly focused on internal, rather than the external, negotiations. It therefore tends to be unprepared for the changing positions of other countries, and hence is unable to react swiftly.[318]

Developing country participation was one of the key US demands, and was regarded as reasonable by many other countries, despite its explicit exclusion from a protocol by the Berlin Mandate. Therefore, the outcome remained unclear until the end of the conference. It had been a highly sensitive issue throughout the Kyoto negotiations, and again evoked heated debate when the issue came up for adoption during the final COW session. Whilst China, India and OPEC were the main voices demanding the deletion of any provision for developing country participation, its inclusion was supported by developed and developing countries alike. The proponents had a numerical majority, making it feasible to adopt some provision for future voluntary participation by developing countries, possibly in the manner that Article 17 on emissions trading was adopted.

However, Estrada had long been opposed to binding obligations from developing countries because they were contrary to the spirit of the Berlin Mandate. Yet, he was aware of the need to establish a long-term process that incorporated developing countries.[319] In the end, he decided to delete the article after the developing countries – including China and India – had accepted the concept of emissions trading two hours earlier. It later allowed the CDM to be adopted. This was a clever trade-off by Estrada. Furthermore, by 6 a.m., with many draft articles remaining to be adopted, it was too late to allow for further discussions.

Both events demonstrate the level of influence the Chair of the COW had on the negotiating process. Unquestionably, he has been one of the most influential individuals in the FCCC process.[320] YAMIN praises his 'affable manner and jocular style of chairing [which] disguised his steely grip over the proceedings and his ability, through offence and charm dispensed in roughly equal measure, to bring matters to closure'.[321]

Estrada's use of the gavel became the most prominent means of decision

318 Gupta and Grijp (forthcoming), *op. cit.*.

319 See, for example, Oberthür and Ott (1999), *op. cit.*, p. 229.

320 Ott (1998), *op. cit.*.

321 Yamin (1998), *op. cit.*, p. 115.

making during the final hours of COP3. Moreover, whenever agreement could not be reached on an issue, he took a decision on behalf of the COW. Anyone opposed to it could formally challenge his ruling and a two-thirds majority could then overturn it. However, no one raised a challenge during the course of the negotiations.[322]

In addition to the Chair's strong leadership, the rules of procedure of the COP created a peculiar situation. At the time of the adoption of the Kyoto Protocol by the COP, the rules of procedure had still not been agreed upon.[323] The main outstanding issue concerned the majorities required for the adoption of specific types of decisions on matters of substance – including the adoption of a protocol – and the views of the parties remained widely divergent.[324] The draft rules were thus applied, with the exception of draft rule 42, on voting.[325] A protocol thus had to be adopted by consensus.

The alternative to a protocol was an amendment to the Convention, which only required a three-quarters majority under Article 15 of the Convention. However, this option was not favoured by any party as it would have undermined the spirit of the Convention's provisions.[326] The status on the voting rules thus granted the Chair significant leeway in judging when there was

---

322 Oberthür and Ott (1999), *op. cit.*, p. 89.

323 The Convention calls for the adoption by consensus of rules of procedure for the COP and its subsidiary bodies (Article 7.2(k), FCCC).

324 They are contained in document FCCC/CP/1997/5.

325 The draft rules are contained in document A/AC.237/L.22/Rev.2, as amended by document FCCC/CP/1995/2. For details on the exception of draft rule 42 see FCCC/CP/1997/5, para.10.

326 Yamin (1998), *op. cit.*, p. 126, n. 16.

enough support to warrant consensus, even when some dissent remained evident.[327]

The Kyoto Protocol had to be adopted as a package, and it remained uncertain until the very end of the conference whether an agreement would be reached. In the end, the extreme pressure to achieve a successful outcome and Chair Estrada's outstanding leadership enabled a consensus to be reached that struck a delicate balance between the widely divergent interests at stake.

327 Grubb *et al.* (1999), *op. cit.*, p. 45, n. 15.

5. CONCLUSION

Meteorology became a priority research area in the 1950s not from scientific findings on climate change and its possible impacts on humanity, but from US governmental and industrial initiatives to promote technological advancement in jet aviation, computers and atomic warfare. This created a global research infrastructure that, after a series of international scientific meetings, came to the conclusion by the late 1970s that the problem of climate change was immediate enough to warrant a political response.

The climate change issue became politicised in the late 1980s, amid growing environmental awareness in industrialised countries, and epitomised by the successful negotiations of the Montreal Protocol in 1987. The first crucial stage in this process was the establishment of a UN Framework Convention on Climate Change in 1992.

The Convention sets as its ultimate objective the stabilisation of greenhouse gas concentrations in the atmosphere at a level that would preserve the global climate system. It commits industrialised countries, albeit in a legally non-binding way, to stabilise their greenhouse gas emissions at 1990 levels by the year 2000, and establishes a set of principles to guide the process. Moreover, the Convention calls on participating states to periodically review its commitments in light of the best available scientific evidence of climate change and its impacts.

However, the political response to climate change has been guided by the national interests of 'critical actors' – the EU, US and Japan – rather than by scientific concerns.

The EU has sought to adopt a leadership role on the issue. Upon unilateral responses by some EC member countries and recognition of the international scope of the problem, it launched a common CO_2 stabilisation target in 1990, and a set of proposals in 1992 to achieve it. Furthermore, it took the lead in pressing for a more substantive Convention. However, the EU has been much criticised for its incapacity to match rhetoric with results. Its common policies have mostly floundered due to institutional deficiencies, and its insistence on conditioning its own action upon like action in other industrialised countries has provoked much criticism.

The EU's strategy for the 1997 Kyoto Conference was to push for a much higher target than it could accomplish, in order to encourage more ambitious numerical targets for the US and Japan. It resisted the flexibility mechanisms, consenting to them in the end only in exchange for provisions limiting their scope. Thus, it secured support for higher targets, the inclusion of its bubble concept, and (weakened) policies and measures designed to enhance the Protocol commitments. The EU's leadership strategy was to set high goals, and then pull other crucial actors toward them.

At the moment, it is rather unlikely that the EU member countries will ratify the Kyoto Protocol without the US. However, some analysts argue that the EU should proceed with ratification. Furthermore, it should demonstrate diplomatic leadership by convincing Japan and Russia, rather than the US, to do likewise.[328]

The US has undergone major shifts in its approach to the climate change issue. It originally started as one of the main promoters for the science. However, with increasing recognition of the potential economic scope of the problem, US industry and some leading figures in the Bush administration encouraged a more laggard position; this emerged most clearly during UNCED in 1992. This laggard position fuelled significant differences between Republicans and industry, on the one side, and Democrats and environmentalists, on the other.

The US strategy for COP3 was to insert in the Protocol as much flexibility as possible, and argue for binding commitments on the part of developing countries. It only succeeded on the first priority, with incorporation of joint implementation, emissions trading and the clean development mechanism.

Given the present political constellation, it is unlikely that US ratification of the Protocol will proceed quickly enough to achieve its target in the first commitment period. Future developments in US climate change policy will depend to a large extent on the outcome of the presidential elections in November 2000 and concurrent changes in the make-up of Congress. Al Gore

---

[328] This 'EU leadership strategy' was developed by Hermann Ott and Sebastian Oberthür, and was firstly presented in public by Hermann Ott on 11 October 1999, in Bonn. 'Fachtagung: Globaler Klimaschutz am Scheideweg'. This is elaborated on in their publication *The Kyoto Protocol, International Climate Change Policy for the New Millennium*.

has a personal political stake in seeing the Kyoto Protocol implemented, and has already promised to submit it to Congress should he become President. Furthermore, the growing acceptance of climate change science in US industry and public opinion has changed the nature of debate on the issue.[329] Thus, there are promising signs that the US will continue to reassess its position as laggard, and contribute more constructively to the climate change process.

It was international pressure that forced Japan in 1990 to develop national climate change policies and emission targets. Despite its impressive record of energy efficiency, Japan is the fourth largest emitter of GHGs. Taking the opportunity to improve its international standing in the non-military arena, Japan has begun asserting its leadership in environmental politics. Moreover, by offering to host COP3, it assumed responsibility for hosting a conference of tremendous significance.

Japan's strategy for COP3 was to occupy a middle position between those of the EU and US – a position it was willing to modify if required. There was not a single issue discussed at Kyoto on which Japan held a view opposing both the EU and US positions at the same time. Japan shared the EU positions on the number of gases to be included in the basket, joint implementation and developing country participation, whilst sharing the US positions on the commitment period, sinks, policies and measures, the EU bubble and the clean development mechanism. It took the middle ground on numerical targets and emissions trading. Japan's ultimate priority lay in bringing the negotiations to a successful conclusion.

Although Japan is willing to ratify and comply with the Protocol, current domestic measures are not expected to achieve more than a stabilisation of 1990 levels by the end of the first commitment period. There is agreement between MITI and the Environment Agency that the Kyoto mechanisms will play an important part in achieving the additional reductions.[330]

The Kyoto Protocol is in many ways a mosaic of the different positions and strategies of the three critical actors. On the one hand, its broad flexibility bears the fingerprints of the US. On the other hand, the inclusion of higher reduction

---

329 These are views expressed by Lisa D. McNeilly, PEW Center on Global Climate Change, Arlington, US, 'Fachtagung: Globaler Klimaschutz am Scheideweg, 11 October, Bonn.

330 This perspective was given by Rie Sugiyama, Institute for Global Environmental Strategies (IGES), Shonan, Japan, 14 October 1999.

commitments in the range of 6-8 per cent of 1990 levels for the EU, US and Japan marks a victory for the EU. However, not all aspects of the Protocol can be deduced purely from national positions or strategies. The clean development mechanism and the exclusion of voluntary participation by developing countries were, in the end, subject to the lateness of the hour, the exhaustion of the delegates and, above all, to the tremendous leadership and insight of Chair Estrada.

REFERENCES

Publications

ANDERSON, J.W. (1998), 'The Kyoto Protocol on Climate Change, Background, Unresolved Issues and Next Steps', Resources for the Future, January 1998, accessible via: http://www.rff.org/about_rff/anderson.htm.

ANDREWS, Richard N. L. (1997), 'United States', pp. 25-43, in: Martin Jänicke and Helmut Weidner (eds.), *National Environmental Policies, A Comparative Study of Capacity-Building*, Berlin.

BARRETT, Brendan F.D. and CHAMBERS, W. Bradnee (1998 Consultation Draft), 'Primer on Scientific Knowledge and Politics in the Evolving Global Climate Regime, COP3 and the Kyoto Protocol', UNU/IAS, Tokyo, June 1998, an extended version of this primer will be published by UNU Press, Tokyo, by late 1999.

BEDER, Sharon (1999), 'Corporate Hijacking of the Greenhouse Debate', pp. 119-122, in: *The Ecologist*, Vol. 29, No. 2, March/April 1999.

BENEDICK, Richard E. (1991), *Ozone Diplomacy*, Cambridge, Mass..

BENEDICK, Richard E. (1998), 'How to Salvage the Kyoto Protocol', Weathervane, March 1998, at: <http://www.weathervane.rff.org/pop/pop5/benedick.html>.

BEUKERING, Pieter van and VELLINGA, Pier (1996), 'Climate change: From science to global policies', pp. 187-215, in: Peter B. Sloep and Andrew Blowers (eds.), *Environmental Policy in an International Context: Environmental Problems as Conflicts of Interest*, London.

BODANSKI, Daniel (1993), 'The United Nations Framework Convention on Climate Change: A Commentary', pp. 451-558, in: *The Yale Journal of International Law*, Vol. 18, No. 2, Summer 1993.

BODANSKI, Daniel (1994), 'Prologue to the Climate Change Convention', pp. 45-74, in: Irving M. Mintzer and J. Amber Leonard (eds.), *Negotiating*

Climate Change: The Inside Story of the Rio Convention, Cambridge.

BRENTON, Tony (1994), *The Greening of Machiavelli, The Evolution of International Environmental Politics*, RIIA, London.

BROWN, Paul (1996), *Global Warming, Can Civilization Survive?*, London.

CAMERON, James (1999), 'The Precautionary Principle', pp. 239-269, in: Gary P. Sampson and W. Bradnee Chambers (eds.), *Trade, Environment and the Millennium*, Tokyo.

CAMPBELL, Karen (1998), 'From Rio to Kyoto: The Use of Voluntary Agreements to Implement the Climate Change Convention', pp. 159-169, in: *Review of European Community & International Environmental Law*, Volume 7, Issue 2, 1998.

CAR, Donald A. and THOMAS, William L. (1998), 'The Kyoto Protocol and US Climate Change Policy: Implications for American Industry', pp. 191-201, in: *Review of European Community & International Environmental Law*, Volume 7, Issue 2, 1998.

COLLIER, Ute (1997), 'The EU and Climate Change Policy: The Struggle over Policy Competences', pp. 43-64, in: Ute Collier and Ragnar E. Löfstedt, *Cases in Climate Change Policy, Political Reality in the European Union*, London.

CONTE GRAND, Mariana (1999), 'Argentina's Voluntary Commitment Proposal at COP5: What is it About?', Weathervane, At the Negotiating Table, December 1999, at: <http://www.weathervane.rff.org/negtable/Argentina's_Proposal.html>.

COOPER, Richard N. (1998), 'Towards a Real Global Warming Treaty', pp. 66-79, in: *Foreign Affairs*, Vol. 77, No. 2, March/April 1998.

CULLET, P. and KAMERI-MBOTE, P. (1998), 'Joint Implementation and Forestry Projects: Conceptual and Operational Fallacies', pp. 393-408, in: *International Affairs*, Vol. 74, No. 2, 1998.

DEPLEDGE, Joanna (no date), 'From Kyoto to Buenos Aires: Prospects for COP4', pp. 1-6, in: W. Bradnee Chambers (ed.), *Global Climate Governance, Inter-linkages between the Kyoto Protocol and other multilateral regimes*, Tokyo.

DESAI, Bharat H. (1999), 'Institutionalizing the Kyoto Climate Accord', pp. 159-163, in: *Environmental Policy and Law*, 29/4 (1999).

DOLZER, Rudolf (1994), 'The Legal Status of the Commitments in the Convention on Climate Change and the Need for Future Revisions', pp. 131-145, in: Ted Hanisch (ed.), *Climate Change and the Agenda for Research*, Boulder.

EHRMANN, Markus and OBERTHÜR, Sebastian (1997), 'Spring in Climate Negotiations?', pp. 192-196, in: *Environmental Policy and Law*, 27/3 (1997).

ESTRADA-Oyuela, Raul (1998 Advance Distribution Copy), 'First Approaches and Unanswered Questions', pp. 18-24, in:José Goldemberg (ed.), Issues & Options, The Clean Development Mechanism, UNDP.

FERMANN, Gunnar (1993), 'Japan's 1990 Climate Policy under Pressure', pp. 287-300, in: *Security Dialogue*, Vol. 24(3), September 1993.

FERMANN, Gunnar (1994), 'Political Leadership and the Development of Problem-solving Capacity in the Global Greenhouse: Prospects for Germany, Japan and the United States Towards the 21st Century', EED Report 1994/3, The Fridtjof Nansen Institute, Lysaker.

FERMANN, Gunnar (1997), 'Political Leadership and Climate Change: The Prospects of Germany, Japan and the United States', pp. 341-405, in: Gunnar Fermann (ed.), *Politics of Climate Change, Key Issues and Critical Actors*, Oslo.

FISCHER, Wolfgang (1992), 'Die Klimapolitik der Staaten zwischen globaler Verantwortung und nationalen Interessen', pp. 164-223, in: P.Borsch and P.M. Wiedemann (eds.), *Was wird aus unserem Klima?*, Landsberg a.L..

FISCHER, Wolfgang and HOLTRUP, Petra (1998a), 'Erfolge und Defizite der internationalen Klimaschutzpolitik', pp. 243-292, in: Peter Borsch and Jürgen-Friedrich Hake (eds.), *Klimaschutz: eine globale Herausforderung*, Landsberg a.L..

FISCHER, Wolfgang and HOLTRUP, Petra (1998b), 'Institutionelle Strukturen und Entscheidungsprozesse der Umweltaußenpolitik', pp. 121-136, in: Wolf-Dieter Eberwein and Karl Kaiser (eds.), *Deutschlands neue Außenpolitik, Band 4: Institutionen und Ressourcen*, Munich.

FRANZ, Wendy E. (1997), 'The Development of an International Agenda for Climate Change: Connecting Science to Policy', ENRP Discussion Paper E-97.07, Kennedy School of Government, Harvard University, Cambridge, Mass..

FUKASAKU, Yukiko (1995), 'Energy and environment policy integration', pp. 1063-1076, in: *Energy Policy*, 1995, Vol. 23, No. 12.

GEHRING, Thomas (1996), 'Das internationale Regime zum Schutz der Ozonschicht: Modell für das Klimaregime', pp. 49-60, in: Hans Günther Brauch (ed.), *Klimapolitik*, Berlin.

GORE, Al (1993), *Earth in the Balance, Ecology and the Human Spirit*, New York.

GÖRRISSEN, Thorsten (1993), *Grenzüberschreitende Umweltprobleme in der internationalen Politik*, Baden-Baden.

GRUBB, Michael (1993), 'United Nations Framework Convention on Climate Change', pp. 61-73, in: Michael Grubb et al., *The Earth Summit Agreements, A Guide and Assessment*, London.

GRUBB, Michael (1998), 'International Emissions Trading under the Kyoto Protocol: Core Issues in Implementation', pp. 140-146, in: *Review of European Community & International Environmental Law*, Volume 7, Issue 2, 1998.

GRUBB, Michael, with VROLIJK, Christiaan and BRACK, Duncan (1999), *The Kyoto Protocol, A Guide and Assessment*, London.

GUERRERO AGUIRRE, Francisco Javier (1995), 'Power, Asymmetry and Negotiation: A Theoretical Analysis', Summer 1995, at: <http://snipe.ukc.ac.uk/international/papers.dir/paco1.html> as of 10 January 1999.

GUPTA, Joyeeta and van der GRIJP, Nicolien (forthcoming), 'Leadership in the Climate Change Regime: The European Union in the Looking Glass', to be published in the *International Journal of Environment and Pollution*.

HAAS, Peter (1992), 'Introduction: Epistemic Communities and International Policy Coordination', pp. 1-35, in: *International Organization* 46,1 (1992).

HAIGH, Nigel (1996), 'Climate Change Policies and Politics in the European

Community', pp. 155-185, in: Jill Jäger and Tim O'Riordan (eds.), *Politics of Climate Change*, London.

HARRIS, Paul G. (1998), 'Understanding America's Climate Change Policy: Realpolitik, Pluralism, and Ethical Norms', OCEES Research Paper No. 15, Oxford.

HARRISON, Neil Edward (1994), 'Heads in the Clouds, Feet in the Sand: Multilateral Policy Coordination in Global Environmental Issues', Dissertation, University of Denver, published by University Microfilms International, Ann Arbor, MI..

HATCH, Michael (1993), 'Domestic Politics and International Negotiations: the Politics of Global Warming in the United States', pp. 1-39, in: *Journal of Environment and Development* 2, 2 (Summer 1993).

HAYES, Peter (1993), 'Constructing a global greenhouse regime', pp. 319-356, in: Peter Hayes and Kirk Smith (eds.), *The global greenhouse regime, Who pays?*, Tokyo.

HELLER, Thomas C. (1999), 'Environmental Realpolitik: Joint Implementation and Climate Change', April 1999, at: <http://www.law.indiana.edu/glsj/vol3/no2/heller.html>.

HORSCH, Richard A. and RICHARDS, Joseph D., 'Does Kyoto Protocol Fall Short of the Mark?', New York Law Journal, at: <http://www.ljx.com/practice/internat/0427kyoto.html> as of 27 April 1998.

HOUGHTON, G.J, JENKINS, G.J. and EPHRAUMS, J.J. (eds.) (1990), Climate Change, The IPCC Scientific Assessment, Cambridge.

HYVARINEN, Joy (1999), 'The European Community's Monitoring Mechanism for CO_2 and other Greenhouse Gases: the Kyoto Protocol and other Recent Developments, pp. 191-197, in: *Review of European Community & International Environmental Law*, Volume 8, Issue 2, 1999.

INOGUCHI, Takashi (1986), 'Japan's Images and Options: Not a Challenger, but a Supporter', pp. 95-119, in: *Journal of Japanese Studies*, Vol. 12, No. 1, Winter 1986.

INTERNATIONAL COUNCIL OF SCIENTIFIC UNIONS (1996), *Understanding Our Planet*, 2nd ed., Paris.

JACOBY, Henry D., PRINN, Ronald G. and SCHMALENSEE, Richard (1998), 'Kyoto's Unfinished Business', pp. 54-66, in: *Foreign Affairs*, Vol. 77, No. 4, July/August 1998.

JÄGER, Jill and O'RIORDAN, Tim (1996), 'The History of Climate Change Science and Politics', pp. 1-31, in: Jill Jäger and Tim O'Riordan (eds.), *Politics of Climate Change*, London.

JOHNSON, Anne K. (1998), 'The Influence of Political Culture on the Formation of Pre-Regime Climate Change Policies in Sweden, Japan and the United States', pp. 233-244, in: *Environmental Values*, Vol. 7, No. 2, 1998.

KEGLEY, Charles W. Jr. and WITTKOPF, Eugene R. (1997), *World Politics, Trends and Transformations*, 6th ed., New York.

KAWASHIMA, Yasuko (1997), 'A Comprehensive Analysis of the Decision-making Process of Developed Countries toward CO_2 Emissions Reduction Targets', pp. 95-126, in: *International Environmental Affairs*, Vol. 9, No. 2, 1997.

KAWASHIMA, Yasuko (2000), 'Nuclear Power and Climate Change: The Current Situation in Japan and a Message to the United States', Weathervane, Perspectives on Policy, February 2000, at: <http://www.weathervane.rff.org/pop/pop9/kawashima.html>.

KANIE, Norichika, 'Domestic Policy and Leadership in Multilateral Diplomacy: The Case of the Netherlands' Kyoto Protocol Negotiation', in: UNU/IAS Environment and Multilateral Diplomacy Series, No. 3, Tokyo.

KRÄGENOW, Timm (1995), *Verhandlungspoker um Klimaschutz*, Freiburg/Br..

LENSCHOW, Andrea (1996), 'Der umweltpolitische Entscheidungsprozeß in der Europäischen Union am Beispiel der Klimapolitik', pp. 89-104, in: Hans Günter Brauch (ed.), *Klimapolitik*, Berlin.

LINCOLN, Edward J. (1993), *Japan's New Global Role*, Washington, D.C..

LOSKE, Reinhard (1996), *Klimapolitik im Spannungsfeld von Kurzzeitinteressen und Langzeiterfordernissen*, Marburg.

LOWI, Theodore J., and GINSBERG, Benjamin (1998), *American Government*, Brief Fifth Edition, New York.

MACRORY, Richard and HESSION, Martin (1996), 'The European Community and Climate Change, The role of law and legal competence', pp. 106-154, in: Jill Jäger and Tim O'Riordan (eds.), *Politics of Climate Change*, London.

MISSFELDT, Fanny (1998), 'Flexibility Mechanisms: Which Path to Take after Kyoto?' pp. 128-139, in: *Review of European Community & International Environmental Law*, Volume 7, Issue 2, 1998.

MOLTKE, Konrad von and RAHMAN, Atiq (1996), 'External Perspectives on Climate Change, A view from the United States and the Third World', pp. 330-345, in: Jill Jäger and Tim O'Riordan (eds.), *Politics of Climate Change*, London.

NATHAN, Richard (1997), 'Japan seeks to bridge split on emissions policy', p. 541, in: *Nature*, Vol. 387, No. 6634, 12 June 1997.

NATIONAL RESEARCH COUNCIL (1998), 'Overview Global Environmental Change: Research Pathways for the Next Decade', Washington, D.C..

NISHIOKA, S. and HARASAWA, H. (eds.) (1998), *Global Warming, The Potential Impact on Japan*, Tokyo.

NITZE, William A. (1994), 'A Failure of Presidential Leadership', pp. 187-200, in: Irving M. Mintzer and J. Amber Leonard (eds.), *Negotiating Climate Change: The Inside Story of the Rio Convention*, Cambridge.

OBERTHÜR, Sebastian and OTT, Hermann E. (1999), The Kyoto Protocol, International Climate Policy for the 21st Century, Berlin.

OKAMATSU, Sozaburo (1992), 'MITI's Centennial Vision of Global Environment and Technology and the Response to Global Warming: Concerning New Earth 21', pp. 335-348, in: Helmar Krupp, *Energy Politics and Schumpeter Dynamics*, Tokyo.

OTT, Hermann (1996), 'Völkerrechtliche Aspekte der Klimarahmenkonvention', pp. 61-74, in: Hans Günter Brauch (ed.), *Klimapolitik*, Berlin.

OTT, Hermann (1997), 'Das internationale Regime zum Schutz des Klimas', pp. 201-218, in: Thomas Gehring and Sebastian Oberthür (eds.), *Internationale Umweltregime, Umweltschutz durch Verhandlungen und Verträge*, Opladen.

OTT, Hermann E. (1998), 'The Kyoto Protocol to the UN Framework

Convention on Climate Change – Finished and Unfinished Business', Wuppertal Institut für Klima, Umwelt, Energie GmbH, February 1998, at: <http://www.wupperinst.org/Publikationen/Kyoto_Protokoll.html>.

PAOLETTO, Glen and SCHROEDER, Heike (1997), 'Enhancing Participation of NGOs in the UNFCCC Process', GEIC Paper Series, October 1997.

PATERSON, Mathew (1996), *Global Warming and Global Politics*, London.

PATERSON, Mathew and GRUBB, Michael (1992), 'The international politics of climate change', pp. 293-311, in: *International Affairs*, 68, 2(1992).

PETERS, Hans-Peter and SIPPEL, Marion (1998), 'Der Treibhauseffekt als journalistische Herausforderung', pp. 293-316, in: PeterBorsch and Jürgen-Friedrich Hake (eds.), *Klimaschutz: eine globale Herausforderung*, Landsberg a.L..

PLATE, Petra A. (1997), 'Japan und das Kohlendioxydproblem vor der Klimakonferenz in Kyoto', pp. 587-593, in: *Japan*, December 1997.

PLATE, Petra A. (1998), 'Ergebnisse der Klimakonferenz in Kyoto: Die Pflicht zum Forschen und Mitdenken', pp. 73-78, in: *JAPAN aktuell*, February 1998.

RINGIUM, Lasse (1997), 'Differentiation, Leaders and Fairness, Negotiating Climate Commitments in the European Community', CICERO Report 1997:8, Oslo.

ROWBOTHAM, Elizabeth J. (1996), 'Legal Obligation and Uncertainties in the Climate Change Convention', pp. 32-50, in: Jill Jäger and Tim O'Riordan (eds.), *Politics of Climate Change: A European Perspective*, London.

SCHREURS, Miranda A. (1996), 'Domestic Institutions, International Agendas, and Global Environmental Protection in Japan and Germany', Dissertation, University of Michigan, published by University Microfilms International, Ann Arbor, MI..

SCHUMER, Sylvia (1996), 'Die Europäische Union als Akteur in der internationalen Umweltpolitik am Beispiel desKlimaregimes', pp. 105-114, in: : Hans Günter Brauch (ed.), *Klimapolitik*, Berlin.

SCHUPPERT, Stefan (1998), *Neue Steuerungselemente im Umweltvölkerrecht am Beispiel des Montrealer Protokolls und des Klimaschutzrahmenüberein-*

kommens, Max-Planck-Institut für ausländisches öffentliches Recht und Völkerrecht, Band 131, Berlin.

SIMONIS, Udo E. (1997), 'Kyoto-Konferenz: Ein Rückblick auf die japanische Position zur internationalen Klimapolitik', pp. 452-456, in: *Zeitschrift für Angewandte Umweltforschung*, Jahrgang 10 (1997), Heft 4.

SIMONIS, Udo E. (1998), 'Das "Kyoto-Protokoll", Aufforderung zu einer innovativen Klimapolitik', Wissenschafttszentrum Berlin für Sozialforschung, FS II 98-403, Berlin.

SJÖSTEDT, Gunnar (1998), 'The EU Negotiates Climate Change, External Performance and Internal Structural Change', pp. 227-256, in: *Cooperation and Conflict, Nordic Journal of International Studies*, Vol. 33, No. 3, September 1998.

SKOLNIKOFF, Eugene B. (1990), 'The Policy Gridlock on Global Warming', pp. 77-93, in: *Foreign Policy* 79, Summer 1990.

SKOLNIKOFF, Eugene B. (1997), 'Same Science, Differing Policies; The Saga of Global Climate Change', Report No. 22, MIT Joint Program on the Science and Policy of Global Change, Cambridge, Mass., August 1997, at: <http://www.mit.edu/globalchange/www/rpt22.html>.

SMITH, Kirk R., SWISHER, Joel and AHUJA, Dilip R., 'Who pays (to solve the problem and how much)?', pp. 70-98, in: Peter Hayes and Kirk Smith (eds.), *The global greenhouse regime, Who pays?*, Tokyo.

SWANN, Dennis (1992), *The Economics of the Common Market*, 7th ed., London.

TAMITATE, Takao (1993), 'Japan's Climate Policy, A Rejoinder', pp. 301-304, in: *Security Dialogue*, Vol. 24(3), September 1993.

TOMAN, Michael and CAZORLA, Marina (1998), 'The Clean Development Mechanism: A Primer', Resources for the Future, September 1998, at: <http://teriin.org/treat/docs/res2.htm>.

UI, Jun (1989), 'Anti-Pollution Movement and Other Grass-Roots Organizations', pp. 109-119, in: Shigeto Tsuru and Helmut Weidner (eds.), *Environmental Policy in Japan*, Berlin.

UNGAR, Sheldon (1999), 'Is Strange Weather in the Air? A Study of U.S: National Network News Coverage of Extreme Weather Events', pp. 133-

150, in: *Climate Change*, Vol. 41, No. 2, February 1999.

UNU, GEIC, IAS (1998), *Global Climate Governance: A Report on the Inter-Linkages between the Kyoto Protocol and other Multilateral Regimes*, Report – Part 1, Tokyo.

UNU, GEIC, IAS (1998), *Global Climate Governance: Scenarios and Options on the Inter-Linkages between the Kyoto Protocol and other Multilateral Regimes*, Report – Part 2, Tokyo.

URWIN, Derek W. (1991), *The Community of Europe, A History of European Integration since 1945*, Harlow.

VERNON, R. (1993), 'Behind the scenes: how policymaking in the EC, Japan and US affects global negotiations, pp. 12-20, 35-42, in: *Environment* 35 (5).

VONKEMANN, Gerrit (1996), 'International co-operation: the European Union', pp. 105-134, in: Andrew Blowers and Pieter Glasbergen, *Environmental Policy in an International Context: Prospects for Environmental Change*, London.

WAGNER, Jay P. (1997), 'The Climate Change Policy of the European Union', pp. 297-340, in: Gunnar Fermann (ed.), *International Politics of Climate Change, Key Issues and Critical Actors*, Oslo.

WALLACE, David (1995), *Environmental Policy and Industrial Innovation, Strategies in Europe, the United States and Japan*, London.

WERKSMAN, Jacob (1998), 'The Clean Development Mechanism: Unwrapping the 'Kyoto Surprise", pp. 147-158, in: *Review of European Community & International Environmental Law*, Volume 7, Issue 2, 1998.

YAMIN, Farhana (1998), 'The Kyoto Protocol: Origins, Assessment and Future Challenges, pp. 113-127, in: *Review of European Community & International Environmental Law*, Volume 7, Issue 2, 1998.

Magazine and Newspaper Articles

Der Spiegel 25/1992, pp. 148-151, 'Umwelt-Gipfel: Der mühevolle Weg zur ökologischen Wende'.

Der Spiegel 14/1995, pp. 36-38, 'Hohepriester im Kohlenstoff-Klub'.

Die tageszeitung, 4 December 1997, Peter Tautfest, 'Die europäische Arroganz'.

Die tageszeitung, 9 December 1997, Bernd Ulrich, 'Vom Saulus zum Paulus zum Saulus'.

DIE ZEIT, 28 November 1997, Hans Schuh, 'Vor dem Klimagipfel: Der Streit der Experten'.

[331]*Earth Negotiations Bulletin*, 10 April 1995, Vol. 12 No. 21.

Earth Negotiations Bulletin, 18 December 1995, Vol. 12 No. 25.

Earth Negotiations Bulletin, 22 July 1996, Vol. 12 No. 38.

Earth Negotiations Bulletin, 1 December 1997, Vol. 12, No. 67 – 10 December 1997, Vol. 12, No. 75.

Earth Negotiations Bulletin, 13 December 1997, Vol. 12, No. 76, Summary Report of COP3.

Earth Negotiations Bulletin, 16 November 1998, Vol. 12, No. 97, Summary Report of COP4.

Earth Negotiations Bulletin, 8 November 1999, Vol. 12, No. 123, Summary Report of COP5.

[332]*ECO*, 1 December 1997, Issue No. 1, Vol. XCVIII – 10 December 1997, Issue No. 10, Vol. XCVIII.

ECO, 11 November 1998, Ed. 5, Vol. C.

Financial Times, 9 December 1997, Leyla Boulton, 'Gore tells US delegates to soften stance on emissions'.

Financial Times, 10 December 1997, Leyla Boulton, 'Global dealers could beat pollution'.

Financial Times, 11 December 1997, Leyla Boulton and Bethan Hutton, 'China and India may delay decision on climate treaty'.

Frankfurter Allgemeine Zeitung, 29 November 1997, Manfred Schäfers, 'In Kyoto soll die Treibjagd auf die Treibhausgase eröffnet werden'.

---

331 *Earth Negotiations Bulletin* can be found on the Internet at: <http://www.iisd.ca/linkages>.

332 *ECO* can be found on the Internet at: <http://www.igc.org/climate/Eco.html>.

Frankfurter Allgemeine Zeitung, 1 December 1997, Anne Schneppen, 'Umweltmittler'.

International Herald Tribune, 1 December 1997, Amy Brown, 'Possible Targets and Timetables'.

International Herald Tribune, 6-7 December 1997, William K. Stevens, 'Pact "Takes Shape" At Warming Talks'.

International Herald Tribune, 9 December 1997, Kevin Sullivan and Joby Warrick, 'Gore Leaves Climate Talks in a Blur'.

International Herald Tribune, 10 December 1997, Joby Warrick and Kevin Sullivan, 'Nearing Finish, Climate Talks Study 2 Offers'.

Japan, December 1997, p. 598, 'Document: Japan's Proposal for Reducing Greenhouse Gases for COP3 Conference in Kyoto Dec. 9-11, 1997'.

Japan, December 1997, pp. 558-559, 'Vorbereitungen zur 3. Internationalen Klimakonferenz in Kyoto: Japan unterbreitet Vertragsentwurf'.

Süddeutsche Zeitung, 8 December 1997, 'EU lehnt unterschiedliche Zielwerte ab'.

Süddeutsche Zeitung 9 December 1997, 'Gore enttäuscht die Europäer'.

The Daily Yomiuri, 3 December 1997, p. 3, Shigeyuki Okajima, 'All must be willing to do with less to curb global warming'.

The Daily Yomiuri, 5 December 1997, Mikiko Miyakawa, 'Japan slams proposal for 'EU bubble'.

The Daily Yomiuri, 7 December 1997, p. 1, Mikiko Miyakawa and Hiromi Sasamoto, 'Kyoto meeting remains divided'.

The Daily Yomiuri, 7 December 1997 p. 3, Hiromi Sasamoto, 'Kyoto climate meeting falters on North-South divide over gas'.

The Daily Yomiuri, 8 December 1997, Mikiko Miyakawa and Justin McCurry, 'EU prepared to accept "differentiated" targets'.

The Daily Yomiuri, 9 December 1997, Justin McCurry, 'Parties inch closer as protocol deadline nears'.

The Daily Yomiuri, 10 December 1997, Mikiko Miyakawa and Hiromi Sasamoto, 'Proposal calls for 5% target'.

The Economist, 21 February 1998, pp. 101-102, 'The earth's climate, Solar rash and earthly fever'.

The Economist, 11 April 1998, p. 81, 'Global warming, In flux'.

The Japan Times, 1 December 1997, Stephen Hesse and Richard N. Mott, 'Pressure on to bridge differences, Strong agreement essential'.

The Japan Times, 2 December 1997, Mitsuko Nashima and Yosuke Naito, 'U.S. hints at compromise'.

The Japan Times, 4 December 1997, Eric Johnston and Yosuke Naito, 'U.S. hints at offering "symbolic" cut'.

The Japan Times, 8 December 1997, p. 1, Mitsuko Nashima and Eric Johnston, 'Kyoto talks enter final leg'.

The Japan Times, 8 December 1997, p. 3, Eric Johnston, 'Kyoto important for Sino-U.S. ties'.

The Japan Times, 9 December 1997, Mitsuko Kashima and Asako Murakami, 'Gore pledges U.S: flexibility'.

The Japan Times, 10 December 1997, Mitsuko Nashima and Eric Johnston, 'Draft targets 5% cut for U.S.'.

The Japan Times, 11 December 1997, Mitsuko Nashima and Eric Johnston, 'Developed nations reach tentative emissions pact'.

Internet Pages

'Denver Summit Countries: Key Economic Data', June 1997, at: <http://www.state.gov/www/issues/economic/summit/fsg7data97.html>.

'European Union – Selected instruments taken from Treaties, Title XVI, Environment', at: <http://europa.eu.int/abc/obj/treaties/en/entr6d16.htm#142> as of 4 May 1999.

'Fifth European Community environment programme: towards sustainability', at: <http://europa.eu.int/scadplus/leg/en/lvb/128062.htm> as of 4 May 1999.

'Japan Environment Quarterly, Vol.2 No.4, November 1997', at: <http://www.eic.or.jp/eanet/e/jeq/v002-04.html> as of 9 July 1999.

'Japan: Positive Leadership from COP3 Chair Required!, ECO Tokyo Special Edition, November 8-9, 1997', at: <http://www.igc.org/climate/ets_jprop.html> as of 24 June 1999.

'Kyoto Protocol Status of Ratification (as at 13 January 2000)', at: <http://www.unfccc.de/resource/kpstats.pdf> as of 18 February 2000.

'President Clinton's Climate Change Proposal, 22 October 1997', at: <http://www.state.gov/www/global/global_issues/climate/background.html > as of 30 October 1998.

'Press Briefing by the Government of Japan, COP3', at: <http://www.mofa.go.jp/policy/global/environment/warm/cop3/session.htm l> as of 9 July 1999.

'Senate Resolution 98, 105th Congress, 1st Session', at: <http://www.senate/gov/~rpc/rva/1051/1051205.htm> as of 1 June 1999.

'Summary for Policymakers: The Science of Climate Change – IPCC Working Group I', at: <http://www.ipcc.ch/pub/sarsum1.htm> as of 8 December 1999.

'US Delegation to COP3, Press Briefing, Kyoto, Japan', at: <http://www.state.gov/www/global/oes> as of 30 October 1998.

<http://www.keidanren.or.jp/english/policy/pol058/index.html>.

<http://www.unfccc.de/fccc/ccinfo/defaij.htm>.

<http://www.worldcorpcom/dc-online/gcc/>.

Documents

A/AC.237/L.22/Rev.2, 'Procedural, Institutional and Legal Matters, Rules of Procedure of the Conference of the Parties and Its Subsidiary Bodies', 15 February 1995.

A/AC.237/L.23, 'Matters Relating to Commitments, Review of the Adequacy of

Commitments in Article 4, Paragraphs 2(a) and (b)', 27 September 1994.

A/AC.237/L.23.Add.1, 'Matters Relating to Commitments, Review of the Adequacy of Commitments in Article 4, Paragraphs 2(a) and (b)' 27 September 1994.

FCCC/AGBM/1996/7, 'Strengthening the Commitments in Article 4.2(a) and (b), Quantified Emission Limitation and Reduction Objectives within Specified Time-Frames, Review of possible indicators to define criteria for differentiation among Annex I Parties', 8 July 1996.

FCCC/AGBM/1997/2, 'Framework Compilation of Proposals from Parties for the Elements of a Protocol or Another Legal Instrument', 31 January 1997.

FCCC/AGBM/1997/INF.2, 'Compilation of Responses from Parties on Issues Related to Sinks', 29 November 1997.

FCCC/AGBM/1997/MISC.4, 'Responses from Parties on Issues Related to Sinks', 18 November 1997.

FCCC/AGBM/1997/MISC.4/Add.1, 'Responses from Parties on Issues Related to Sinks', 24 November 1997.

FCCC/CP/1995/2, 'Adoption of Rules of Procedures', 10 March 1995.

FCCC/CP/1995/7, 'Report of the Conference of the Parties on its first session, held at Berlin from 28 March to 7 April 1995', 24 May 1995

FCCC/CP/1995/7/Add.1, 'Report of the Conference of the Parties on its first session, held at Berlin from 28 March to 7 April 1995, Addendum', 6 June 1995.

FCCC/CP/1996/15/Add.1, 'Report of the Conference of the Parties on its second session, held at Geneva from 8 to 19 July 1995, Addendum', 19 October 1997.

FCCC/CP/1997/2, 'Adoption of a protocol or another legal instrument: Fulfilment of the Berlin Mandate, Revised text under negotiation', 12 November 1997.

FCCC/CP/1997/5, 'Organizational Matters, Adoption of the Rules of Procedure', 19 November 1997.

FCCC/CP/1997/7/Add.1, 'Report of the Conference of the Parties on its third session, held at Kyoto from 1 to 11 December 1997, Addendum', 25 March

1998.

FCCC/CP/1997/CRP.6, 'Kyoto Protocol to the UNFCCC, Final draft by the Chairman of the Committee of the Whole', 10 December 1997.

FCCC/CP/1998/16/Add.1, 'Report of the Conference of the Parties on its fourth session, held at Buenos Aires from 2 to 14 November 1998, Addendum', 25 January 1999.

FCCC/CP/1999/6, 'Report of the Conference of the Parties on its fifth session, held at Bonn from 25 October to 5 November 1999', 21 December 1999.

FCCC/IDR.1/USA, 'United States, Report on the in-depth review of the national communication of the United States of America', 26 February 1996.

FCCC/NC/7, 'Executive summary of the national communications of the United States of America', 25 July 1995.

FCCC/SBSTA/1998/INF.3, 'Activities implemented jointly under the pilot phase', 20 May 1998.

FCCC, *Who is Who in the UNFCCC Process 1996-1997*, 2nd Volume, Bonn.

Framework Convention on Climate Change (1992), UNEP/IUC.

The Kyoto Protocol to the Convention on Climate Change (1998), UNEP/IUC.

Other Sources

Audio recordings of COP3 sessions, UNFCCC Secretariat Library, Bonn.

Carruthers, Campion, unpublished translation of quote from Niccolò Machiavelli, *The Prince*, September 1999.

Depledge, Joanna, personal communication, 26 January 2000.

McNeilly, Lisa D., PEW Center on Global Climate Change, Arlington, US, 'Fachtagung: Globaler Klimaschutz am Scheideweg', Bonn, 11 October 1999.

Remarks by Vice President Al Gore at the COP3 high-level segment, 8 December 1997, on file with author.

Ott, Hermann, Wuppertal Institute on Climate, Environment and Energy, 'Fachtagung: Globaler Klimaschutz am Scheideweg', Bonn, 11 October 1999.

Sugiyama, Rie, Institute for Global Environmental Strategies, Shonan, Japan, Interview, 14 October 1999.

ANNEX

Glossary

AI	Annex I
AGBM	Ad-hoc Group on the Berlin Mandate
AOSIS	Alliance of Small Island States
CCAP	Climate Change Action Plan (US)
CDM	Clean development mechanism
CEITs	Countries with economies in transition
CERs	Certified emission reductions (CDM)
CFCs	Chlorofluorocarbons
CH_4	Methane
CO_2	Carbon dioxide
COP	Conference of the Parties
COP/MOP	Session of the Conference of the Parties serving as the meeting of the Parties to this Protocol
DCs	Developing countries
EC	European Community
EPA	Environmental Protection Agency (US)
ERUs	Emission reduction units (JI)
EU	European Union
FCCC	Framework Convention on Climate Change
G77	Group of 77
GCC	Global Climate Coalition (US)
GEIC	Global Environment Information Centre, Japan

GEF	Global Environment Facility
GHGs	Greenhouse gases
GWP	Global Warming Potential
HFCs	Hydrofluorocarbons
IAS	Institute for Advanced Studies (UNU, Japan)
ICSU	International Council of Scientific Unions
IHDP	International Human Dimensions Programme on Global Environmental Change
IGO	Inter-governmental organisation
IMO	International Meteorological Organisation
INC	Intergovernmental Negotiating Committee on Climate Change
IPCC	Intergovernmental Panel on Climate Change
JI	Joint implementation
LDP	Liberal Democratic Party (Japan)
LUCF	Land use change and forestry
MITI	Ministry of International Trade and Industry (Japan)
MT	Maastricht Treaty
NASA	National Aeronautics and Space Administration
NGO	Non-governmental organisation
N_2O	Nitrous oxide
OECD	Organisation for Economic Cooperation and Development
PAMs	Policies and measures
PFCs	Perfluorocarbons
QELRCs	Quantified emission limitation and reduction commitments
QELROs	Quantified emissions limitations and reductions objectives
SAR	Second Assessment Report (IPCC)
SBI	Subsidiary Body for Implementation
SBSTA	Subsidiary Body for Scientific and Technological Advice

SEA	Single European Act
SF_6	Sulphur hexafluoide
UNCED	United Nations Conference on Environment and Development
UNEP	United Nations Environment Programme
UNFCCC	United Nations Framework Convention on Climate Change
UNU	United Nations University
WMO	World Meteorological Organisation
WTO	World Trade Organisation

Chronology of meetings

Intergovernmental Negotiating Committee meetings

INC-1	4-14 February 1991	Chantilly, US
INC-2	19-28 June 1991	Geneva
INC-3	9-20 September 1991	Nairobi
INC-4	9-20 December 1991	Geneva
INC-5 (2 parts)	18-25 February 1992	New York
	30 April – 9 May 1992	New York
INC-6	7-10 December 1992	Geneva
INC-7	15-20 March 1993	New York
INC-8	16-27 August 1993	Geneva
INC-9	7-18 February 1994	Geneva
INC-10	22 Aug. – 2 Sep. 1994	Geneva
INC-11	6-17 February 1995	New York

Conferences of the Parties

COP 1	28 March – 7 April 1995	Berlin
COP 2	8-19 July 1996	Geneva
COP 3	1-11 December 1997	Kyoto
COP 4	2-13 November 1998	Buenos Aires
COP 5	25 Oct. – 5 Nov. 1999	Bonn

Ad-hoc Group on the Berlin Mandate meetings

AGBM-1	21-25 August 1995	Geneva
AGBM-2	30 Oct. – 3 Nov. 1995	Geneva
AGBM-3	5-8 March 1996	Geneva
AGBM-4	8-19 July 1996	Geneva
AGBM-5	9-13 December 1996	Geneva
AGBM-6	3-7 March 1997	Bonn
AGBM-7	31 July – 7 August 1997	Bonn
AGBM-8	22-31 October 1997	Bonn
AGBM-8 (res.)	30 November 1997	Kyoto

Sessions of the Subsidiary Body for Scientific and Technological Advice

SBSTA-1	28 August – 1 Sep. 1995	Geneva
SBSTA-2	27 Feb. – 4 March 1996	Geneva
SBSTA-3	9-16 July 1996	Geneva
SBSTA-4	16-18 December 1996	Geneva
SBSTA-5	24-28 February 1997	Bonn
SBSTA-6	28 July – 7 August 1997	Bonn
SBSTA-7	20-29 October 1997	Bonn
SBSTA-8	2-12 June 1998	Bonn
SBSTA-9	2-10 November 1998	Buenos Aires
SBSTA-10	31 May – 11 June 1999	Bonn
SBSTA-11	25 Oct. – 5 Nov. 1999	Bonn

Sessions of the Subsidiary Body for Implementation

SBI-1	31 Aug. – 1 Sep. 1995	Geneva
SBI-2	27 Feb. – 4 March 1996	Geneva
SBI-3	9-16 July 1996	Geneva
SBI-4	10-11 December 1996	Geneva
SBI-5	24-28 February 1997	Bonn
SBI-6	28 July – 7 August 1997	Bonn
SBI-7	20-29 October 1997	Bonn
SBI-8	2-12 June 1998	Bonn
SBI-9	3-10 November 1998	Buenos Aires
SBI-10	31 May – 11 June 1999	Bonn
SBI-11	25 Oct. – 25 Nov. 1999	Bonn

United Nations Framework Convention on Climate Change

The Parties to this Convention,

Acknowledging that change in the Earth's climate and its adverse effects are a common concern of humankind,

Concerned that human activities have been substantially increasing the atmospheric concentrations of greenhouse gases, that these increases enhance the natural greenhouse effect, and that this will result on average in an additional warming of the Earth's surface and atmosphere and may adversely affect natural ecosystems and humankind,

Noting that the largest share of historical and current global emissions of greenhouse gases has originated in developed countries, that per capita emissions in developing countries are still relatively low and that the share of global emissions originating in developing countries will grow to meet their social and development needs,

Aware of the role and importance in terrestrial and marine ecosystems of sinks and reservoirs of greenhouse gases,

Noting that there are many uncertainties in predictions of climate change, particularly with regard to the timing, magnitude and regional patterns thereof,

Acknowledging that the global nature of climate change calls for the widest possible cooperation by all countries and their participation in an effective and appropriate international response, in accordance with their common but differentiated responsibilities and respective capabilities and their social and economic conditions,

Recalling the pertinent provisions of the Declaration of the United Nations Conference on the Human Environment, adopted at Stockholm on 16 June 1972,

Recalling also that States have, in accordance with the Charter of the United Nations and the principles of international law, the sovereign right to exploit their own resources pursuant to their own environmental and developmental policies, and the responsibility to ensure that activities within their jurisdiction or control do not cause damage to the environment of other States or of areas beyond the limits of national jurisdiction,

Reaffirming the principle of sovereignty of States in international cooperation to address climate change,

Recognizing that States should enact effective environmental legislation, that environmental standards, management objectives and priorities should reflect the environmental and developmental context to which they apply, and that standards applied by some countries may be inappropriate and of unwarranted economic and social cost to other countries, in particular developing countries,

Recalling the provisions of General Assembly resolution 44/228 of 22 December 1989 on the United Nations Conference on Environment and Development, and resolutions 43/53 of 6 December 1988, 44/207 of 22 December 1989, 45/212 of 21 December 1990 and 46/169 of 19 December 1991 on protection of global climate for present and future generations of mankind,

Recalling also the provisions of General Assembly resolution 44/206 of 22 December 1989 on the possible adverse effects of sea level rise on islands and coastal areas, particularly low- lying coastal areas and the pertinent provisions of General Assembly resolution 44/172 of 19 December 1989 on the implementation of the Plan of Action to Combat Desertification,

Recalling further the Vienna Convention for the Protection of the Ozone Layer, 1985, and the Montreal Protocol on Substances that Deplete the Ozone Layer, 1987, as adjusted and amended on 29 June 1990,

Noting the Ministerial Declaration of the Second World Climate Conference adopted on 7 November 1990,

Conscious of the valuable analytical work being conducted by many States on climate change and of the important contributions of the World Meteorological Organization, the United Nations Environment Programme and other organs, organizations and bodies of the United Nations system, as well as other international and intergovernmental bodies, to the exchange of results of scientific research and the coordination of research,

Recognizing that steps required to understand and address climate change will be environmentally, socially and economically most effective if they are based on relevant scientific, technical and economic considerations and continually re-evaluated in the light of new findings in these areas,

Recognizing that various actions to address climate change can be justified economically in their own right and can also help in solving other environmental problems,

Recognizing also the need for developed countries to take immediate action in a flexible manner on the basis of clear priorities, as a first step towards comprehensive response strategies at the global, national and, where agreed, regional levels that take into account all greenhouse gases, with due consideration of their relative contributions to the enhancement of the greenhouse effect,

Recognizing further that low-lying and other small island countries, countries with low-lying coastal, arid and semi-arid areas or areas liable to floods, drought and desertification, and developing countries with fragile mountainous ecosystems are particularly vulnerable to the adverse effects of climate change,

Recognizing the special difficulties of those countries, especially developing countries, whose economies are particularly dependent on fossil fuel production, use and exportation, as a consequence of action taken on limiting greenhouse gas emissions,

Affirming that responses to climate change should be coordinated with social and economic development in an integrated manner with a view to avoiding adverse impacts on the latter, taking into full account the legitimate priority needs of developing countries for the achievement of sustained economic growth and the eradication of poverty,

Recognizing that all countries, especially developing countries, need access to resources required to achieve sustainable social and economic development and that, in order for developing countries to progress towards that goal, their energy consumption will need to grow taking into account the possibilities for achieving greater energy efficiency and for controlling greenhouse gas emissions in general, including through the application of new technologies on terms which make such an application economically and socially beneficial,

Determined to protect the climate system for present and future generations,

Have agreed as follows:

Article 1
Definitions[333]

For the purposes of this Convention:

1. "Adverse effects of climate change" means changes in the physical environment or biota resulting from climate change which have significant deleterious effects on the composition, resilience or productivity of natural and managed ecosystems or on the operation of socio-economic systems or on human health and welfare.

2. "Climate change" means a change of climate which is attributed directly or indirectly to human activity that alters the composition of the global atmosphere and which is in addition to natural climate variability observed over comparable time periods.

[333] Titles of articles are included solely to assist the reader.

3. "Climate system" means the totality of the atmosphere, hydrosphere, biosphere and geosphere and their interactions.

4. "Emissions" means the release of greenhouse gases and/or theirprecursors into the atmosphere over a specified area and period of time.

5. "Greenhouse gases" means those gaseous constituents of the atmosphere, both natural and anthropogenic, that absorb and re-emit infrared radiation.

6. "Regional economic integration organization" means an organization constituted by sovereign States of a given region which has competence in respect of matters governed by this Convention or its protocols and has been duly authorized, in accordance with its internal procedures, to sign, ratify, accept, approve or accede to the instruments concerned.

7. "Reservoir" means a component or components of the climate system where a greenhouse gas or a precursor of a greenhouse gas is stored.

8. "Sink" means any process, activity or mechanism which removes a greenhouse gas, an aerosol or a precursor of a greenhouse gas from the atmosphere.

9. "Source" means any process or activity which releases a greenhouse gas, an aerosol or a precursor of a greenhouse gas into the atmosphere.

Article 2
Objective

The ultimate objective of this Convention and any related legal instruments that the Conference of the Parties may adopt is to achieve, in accordance with the relevant provisions of the Convention, stabilization of greenhouse gas concentrations in the atmosphere at a level that would prevent dangerous anthropogenic interference with the climate system. Such a level should be achieved within a time frame sufficient to allow ecosystems to adapt naturally to climate change, to ensure that food production is not threatened and to enable economic development to proceed in a sustainable manner.

Article 3
Principles

In their actions to achieve the objective of the Convention and to implement its provisions, the Parties shall be guided, inter alia, by the following:

1. The Parties should protect the climate system for the benefit of present and future generations of humankind, on the basis of equity and in accordance with their common but differentiated responsibilities and respective capabilities. Accordingly, the developed country Parties should take the lead in combating climate change and the adverse effects thereof.

2. The specific needs and special circumstances of developing country Parties, especially those that are particularly vulnerable to the adverse effects of climate change, and of those Parties, especially developing country Parties, that would have to bear a disproportionate or abnormal burden under the Convention, should be given full consideration.

3. The Parties should take precautionary measures to anticipate, prevent or minimize the causes of climate change and mitigate its adverse effects. Where there are threats of serious or irreversible damage, lack of full scientific certainty should not be used as a reason for postponing such measures, taking into account that policies and measures to deal with climate change should be cost-effective so as to ensure global benefits at the lowest possible cost. To achieve this, such policies and measures should take into account different socio-economic contexts, be comprehensive, cover all relevant sources, sinks and reservoirs of greenhouse gases and adaptation, and comprise all economic sectors. Efforts to address climate change may be carried out cooperatively by interested Parties.

4. The Parties have a right to, and should, promote sustainable development. Policies and measures to protect the climate system against human-induced

change should be appropriate for the specific conditions of each Party and should be integrated with national development programmes, taking into account that economic development is essential for adopting measures to address climate change.

5. The Parties should cooperate to promote a supportive and open international economic system that would lead to sustainable economic growth and development in all Parties, particularly developing country Parties, thus enabling them better to address the problems of climate change. Measures taken to combat climate change, including unilateral ones, should not constitute a means of arbitrary or unjustifiable discrimination or a disguised restriction on international trade.

Article 4
Commitments

1. All Parties, taking into account their common but differentiated responsibilities and their specific national and regional development priorities, objectives and circumstances, shall:

(a) Develop, periodically update, publish and make available to the Conference of the Parties, in accordance with Article 12, national inventories of anthropogenic emissions by sources and removals by sinks of all greenhouse gases not controlled by the Montreal Protocol, using comparable methodologies to be agreed upon by the Conference of the Parties;

(b) Formulate, implement, publish and regularly update national and, where appropriate, regional programmes containing measures to mitigate climate change by addressing anthropogenic emissions by sources and removals by sinks of all greenhouse gases not controlled by the Montreal Protocol, and measures to facilitate adequate adaptation to climate change;

(c) Promote and cooperate in the development, application and diffusion, including transfer, of technologies, practices and processes that control, reduce or prevent anthropogenic emissions of greenhouse gases not controlled by the Montreal Protocol in all relevant sectors, including the energy, transport, industry, agriculture, forestry and waste management sectors;

(d) Promote sustainable management, and promote andcooperate in the conservation and enhancement, as appropriate, of sinks and reservoirs of all greenhouse gases not controlled by the Montreal Protocol, including biomass, forests and oceans as well as other terrestrial, coastal and marine ecosystems;

(e) Cooperate in preparing for adaptation to the impacts of climate change; develop and elaborate appropriate and integrated plans for coastal zone management, water resources and agriculture, and for the protection and rehabilitation of areas, particularly in Africa, affected by drought and desertification, as well as floods;

(f) Take climate change considerations into account, to the extent feasible, in their relevant social, economic and environmental policies and actions, and employ appropriate methods, for example impact assessments, formulated and determined nationally, with a view to minimizing adverse effects on the economy, on public health and on the quality of the environment, of projects or measures undertaken by them to mitigate or adapt to climate change;

(g) Promote and cooperate in scientific, technological, technical, socio-economic and other research, systematic observation and development of data archives related to the climate system and intended to further the understanding and to reduce or eliminate the remaining uncertainties regarding the causes, effects, magnitude and timing of climate change and the economic and social consequences of various response strategies;

(h) Promote and cooperate in the full, open and prompt exchange of relevant scientific, technological, technical, socio-economic and legal information related to the climate system and climate change, and to the economic and social consequences of various response strategies;

(i) Promote and cooperate in education, training and public awareness related to climate change and encourage the widest participation in this process, including that of non-governmental organizations; and

(j) Communicate to the Conference of the Parties information related to implementation, in accordance with Article 12.

2. The developed country Parties and other Parties included in annex I commit themselves specifically as provided for in the following:

(a) Each of these Parties shall adopt national policies and take corresponding measures on the mitigation of climate change, by limiting its anthropogenic emissions of greenhouse gases and protecting and enhancing its greenhouse gas sinks and reservoirs. These policies and measures will demonstrate that developed countries are taking the lead in modifying longer-term trends in anthropogenic emissions consistent with the objective of the Convention, recognizing that the return by the end of the present decade to earlier levels of anthropogenic emissions of carbon dioxide and other greenhouse gases not controlled by the Montreal Protocol would contribute to such modification, and taking into account the differences in these Parties' starting points and approaches, economic structures and resource bases, the need to maintain strong and sustainable economic growth, available technologies and other individual circumstances, as well as the need for equitable and appropriate contributions by each of these Parties to the global effort regarding that objective. These Parties may implement such policies and measures jointly with other Parties and may assist other Parties in contributing to the achievement of the objective of the Convention and, in particular, that of this subparagraph;

(b) In order to promote progress to this end, each of these Parties shall communicate, within six months of the entry into force of the Convention for it and periodically thereafter, and in accordance with Article 12, detailed information on its policies and measures referred to in subparagraph (a) above, as well as on its resulting projected anthropogenic emissions by sources and removals by sinks of greenhouse gases not controlled by the Montreal Protocol for the period referred to in subparagraph (a), with the aim of returning individually or jointly to their 1990 levels these anthropogenic emissions of carbon dioxide and other greenhouse gases not controlled by the Montreal Protocol. This information will be reviewed by the Conference of the Parties, at its first session and periodically thereafter, in accordance with Article 7;

(c) Calculations of emissions by sources and removals by sinks of greenhouse gases for the purposes of subparagraph (b) above should take into account the best available scientific knowledge, including of the effective capacity of sinks and the respective contributions of such gases to climate change. The Conference of the Parties shall consider and agree on methodologies for these calculations at its first session and review them regularly thereafter;

(d) The Conference of the Parties shall, at its first session, review the adequacy of subparagraphs (a) and (b) above. Such review shall be carried out in the light of the best available scientific information and assessment on climate change and its impacts, as well as relevant technical, social and economic information. Based on this review, the Conference of the Parties shall take appropriate action, which may include the adoption of amendments to the commitments in subparagraphs (a) and (b) above. The Conference of the Parties, at its first session, shall also take decisions regarding criteria for joint implementation as indicated in subparagraph (a) above. A second review of subparagraphs (a) and (b) shall take place not later than 31 December 1998, and thereafter at regular intervals determined by the Conference of the Parties, until the objective of the Convention is met;

(e) Each of these Parties shall:

(i) Coordinate as appropriate with other such Parties, relevant economic and administrative instruments developed to achieve the objective of the Convention; and

(ii) Identify and periodically review its own policies and practices which encourage activities that lead to greater levels of anthropogenic emissions of greenhouse gases not controlled by the Montreal Protocol than would otherwise occur;

(f) The Conference of the Parties shall review, not later than 31 December 1998, available information with a view to taking decisions regarding such amendments to the lists in annexes I and II as may be appropriate, with the approval of the Party concerned;

(g) Any Party not included in Annex I may, in its instrument of ratification, acceptance, approval or accession, or at any time thereafter, notify the Depositary that it intends to be bound by subparagraphs (a) and (b) above. The Depositary shall inform the other signatories and Parties of any such notification.

3. The developed country Parties and other developed Parties included in annex II shall provide new and additional financial resources to meet the agreed full costs incurred by developing country Parties in complying with their obligations under Article 12, paragraph 1. They shall also provide such financial resources, including for the transfer of technology, needed by the developing country Parties to meet the agreed full incremental costs of implementing measures that are covered by paragraph 1 of this Article and that are agreed between a developing country Party and the international entity or entities referred to in Article 11, in accordance with that Article. The implementation of

these commitments shall take into account the need for adequacy and predictability in the flow of funds and the importance of appropriate burden sharing among the developed country Parties.

4. The developed country Parties and other developed Parties included in annex II shall also assist the developing country Parties that are particularly vulnerable to the adverse effects of climate change in meeting costs of adaptation to those adverse effects.

5. The developed country Parties and other developed Parties included in annex II shall take all practicable steps to promote, facilitate and finance, as appropriate, the transfer of, or access to, environmentally sound technologies and know-how to other Parties, particularly developing country Parties, to enable them to implement the provisions of the Convention. In this process, the developed country Parties shall support the development and enhancement of endogenous capacities and technologies of developing country Parties. Other Parties and organizations in a position to do so may also assist in facilitating the transfer of such technologies.

6. In the implementation of their commitments under paragraph 2 above, a certain degree of flexibility shall be allowed by the Conference of the Parties to the Parties included in Annex I undergoing the process of transition to a market economy, in order to enhance the ability of these Parties to address climate change, including with regard to the historical level ofanthropogenic emissions of greenhouse gases not controlled by the Montreal Protocol chosen as a reference.

7. The extent to which developing country Parties will effectively implement their commitments under the Convention will depend on the effective implementation by developed country Parties of their commitments under the Convention related to financial resources and transfer of technology and will take fully into account that economic and social development and poverty eradication are the first and overriding priorities of the developing country Parties.

8. In the implementation of the commitments in this Article, the Parties shall give full consideration to what actions are necessary under the Convention, including actions related to funding, insurance and the transfer of technology, to meet the specific needs and concerns of developing country Parties arising from the adverse effects of climate change and/or the impact of the implementation of response measures, especially on:

(a) Small island countries;

(b) Countries with low-lying coastal areas;

(c) Countries with arid and semi-arid areas, forested areas and areas liable to forest decay;

(d) Countries with areas prone to natural disasters;

(e) Countries with areas liable to drought and desertification;

(f) Countries with areas of high urban atmospheric pollution;

(g) Countries with areas with fragile ecosystems, including mountainous ecosystems;

(h) Countries whose economies are highly dependent on income generated from the production, processing and export, and/or on consumption of fossil fuels and associated energy-intensive products; and

(i) Land-locked and transit countries.

Further, the Conference of the Parties may take actions, as appropriate, with respect to this paragraph.

9. The Parties shall take full account of the specific needs and special situations of the least developed countries in their actions with regard to funding and transfer of technology.

10. The Parties shall, in accordance with Article 10 take into consideration in the implementation of the commitments of the Convention the situation of Parties, particularly developing country Parties, with economies that are vulnerable to the adverse effects of the implementation of measures to respond to climate change. This applies notably to Parties with economies that are highly dependent on income generated from the production, processing and export, and/or consumption of fossil fuels and associated energy-intensive products and/or the use of fossil fuels for which such Parties have serious difficulties in switching to alternatives.

Article 5
Research and Systematic Observation

In carrying out their commitments under Article 4 , paragraph 1(g), the Parties shall:

(a) Support and further develop, as appropriate, international and intergovernmental programmes and networks or organizations aimed at defining, conducting, assessing and financing research, data collection and systematic observation, taking into account the need to minimize duplication of effort;

(b) Support international and intergovernmental efforts to strengthen systematic observation and national scientific and technical research capacities and capabilities, particularly in developing countries, and to promote access to, and the exchange of, data and analyses thereof obtained from areas beyond national jurisdiction; and

(c) Take into account the particular concerns and needs of developing countries and cooperate in improving their endogenous capacities and capabilities to participate in the efforts referred to in subparagraphs (a) and (b) above.

Article 6
Education, Training and Public Awareness

In carrying out their commitments under Article 4, paragraph 1(i), the Parties shall:

(a) Promote and facilitate at the national and, as appropriate, subregional and regional levels, and in accordance with national laws and regulations, and within their respective capacities:

(i) the development and implementation of educational and public awareness programmes on climate change and its effects;

(ii) public access to information on climate change and its effects;

(iii) public participation in addressing climate change and its effects and developing adequate responses; and

(iv) training of scientific, technical and managerial personnel.

(b) Cooperate in and promote, at the international level, and, where appropriate, using existing bodies:

(i) the development and exchange of educational and public awareness material on climate change and its effects; and

(ii) the development and implementation of education and training programmes, including the strengthening of national institutions and the exchange or secondment of personnel to train experts in this field, in particular for developing countries.

Article 7
Conference of the Parties

1. A Conference of the Parties is hereby established.

2. The Conference of the Parties, as the supreme body of this Convention, shall keep under regular review the implementation of the Convention and any related legal instruments that the Conference of the Parties may adopt, and shall make, within its mandate, the decisions necessary to promote the effective implementation of the Convention. To this end, it shall:

(a) Periodically examine the obligations of the Parties and the institutional arrangements under the Convention, in the light of the objective of the Convention, the experience gained in its implementation and the evolution of scientific and technological knowledge;

(b) Promote and facilitate the exchange of information on measures adopted by the Parties to address climate change and its effects, taking into account the differing circumstances, responsibilities and capabilities of the Parties and their respective commitments under the Convention;

(c) Facilitate, at the request of two or more Parties, the coordination of measures adopted by them to address climate change and its effects, taking into account the differing circumstances, responsibilities and capabilities of the Parties and their respective commitments under the Convention.

(d) Promote and guide, in accordance with the objective and provisions of the Convention, the development and periodic refinement of comparable methodologies, to be agreed on by the Conference of the Parties, inter alia, for preparing inventories of greenhouse gas emissions by sources and removals by sinks, and for evaluating the effectiveness of measures to limit the emissions and enhance the removals of these gases;

(e) Assess, on the basis of all information made available to it in accordance with the provisions of the Convention, the implementation of the Convention by the Parties, the overall effects of the measures taken pursuant to the Convention, in particular environmental, economic and social effects as well as their cumulative impacts and the extent to which progress towards the objective of the Convention is being achieved;

(f) Consider and adopt regular reports on the implementation of the Convention and ensure their publication;

(g) Make recommendations on any matters necessary for the implementation of the Convention;

(h) Seek to mobilize financial resources in accordance with Article 4, paragraphs 3, 4 and 5, and Article 11;

(i) Establish such subsidiary bodies as are deemed necessary for the implementation of the Convention;

(j) Review reports submitted by its subsidiary bodies and provide guidance to them;

(k) Agree upon and adopt, by consensus, rules of procedure and financial rules for itself and for any subsidiary bodies;

(l) Seek and utilize, where appropriate, the services and cooperation of, and information provided by, competent international organizations and intergovernmental and non-governmental bodies; and

(m) Exercise such other functions as are required for the achievement of the objective of the Convention as well as all other functions assigned to it under the Convention.

3. The Conference of the Parties shall, at its first session, adopt its own rules of procedure as well as those of the subsidiary bodies established by the Convention, which shall include decision-making procedures for matters not already covered by decision- making procedures stipulated in the Convention. Such procedures may include specified majorities required for the adoption of particular decisions.

4. The first session of the Conference of the Parties shall be convened by the interim secretariat referred to in Article 21 and shall take place not later than one year after the date of entry into force of the Convention. Thereafter, ordinary sessions of the Conference of the Parties shall be held every year unless otherwise decided by the Conference of the Parties.

5. Extraordinary sessions of the Conference of the Parties shall be held at such other times as may be deemed necessary by the Conference, or at the written request of any Party, provided that, within six months of the request being communicated to the Parties by the secretariat, it is supported by at least one-third of the Parties.

6. The United Nations, its specialized agencies and the International Atomic Energy Agency, as well as any State member thereof or observers thereto not Party to the Convention, may be represented at sessions of the Conference of the Parties as observers. Any body or agency, whether national or international, governmental or non-governmental, which is qualified in matters covered by the Convention, and which has informed the secretariat of its wish to be represented at a session of the Conference of the Parties as an observer, may be so admitted unless at least one-third of the Parties present object. The admission and participation of observers shall be subject to the rules of procedure adopted by the Conference of the Parties.

Article 8
Secretariat

1. A secretariat is hereby established.

2. The functions of the secretariat shall be:

(a) To make arrangements for sessions of the Conference of the Parties and its subsidiary bodies established under the Convention and to provide them with services as required;

(b) To compile and transmit reports submitted to it;

(c) To facilitate assistance to the Parties, particularly developing country Parties, on request, in the compilation and communication of information required in accordance with the provisions of the Convention;

(d) To prepare reports on its activities and present them to the Conference of the Parties;

(e) To ensure the necessary coordination with the secretariats of other relevant international bodies;

(f) To enter, under the overall guidance of the Conference of the Parties, into such administrative and contractual arrangements as may be required for the effective discharge of its functions; and

(g) To perform the other secretariat functions specified in the Convention and in any of its protocols and such other functions as may be determined by the Conference of the Parties.

3. The Conference of the Parties, at its first session, shall designate a permanent secretariat and make arrangements for its functioning.

Article 9
Subsidiary Body for Scientific and Technological Advice

1. A subsidiary body for scientific and technological advice is hereby established to provide the Conference of the Parties and, as appropriate, its other

subsidiary bodies with timely information and advice on scientific and technological matters relating to the Convention. This body shall be open to participation by all Parties and shall be multidisciplinary. It shall comprise government representatives competent in the relevant field of expertise. It shall report regularly to the Conference of the Parties on all aspects of its work.

2. Under the guidance of the Conference of the Parties, and drawing upon existing
competent international bodies, this body shall:

(a) Provide assessments of the state of scientific knowledge relating to climate change and its effects;

(b) Prepare scientific assessments on the effects of measures taken in the implementation of the Convention;

(c) Identify innovative, efficient and state- of-the-art technologies and know-how and advise on the ways and means of promoting development and/or transferring such technologies;

(d) Provide advice on scientific programmes, internationalcooperation in research and development related to climate change, as well as on ways and means of supporting endogenous capacity-building in developing countries; and

(e) Respond to scientific, technological and methodological questions that the Conference of the Parties and its subsidiary bodies may put to the body.

3. The functions and terms of reference of this body may be further elaborated by the Conference of the Parties.

Article 10
Subsidiary Body for Implementation

1. A subsidiary body for implementation is hereby established to assist the Conference of the Parties in the assessment and review of the effective implementation of the Convention. This body shall be open to participation by all Parties and comprise government representatives who are experts on matters related to climate change. It shall report regularly to the Conference of the Parties on all aspects of its work.

2. Under the guidance of the Conference of the Parties, this body shall:

(a) Consider the information communicated in accordance with Article12, paragraph 1, to assess the overall aggregated effect of the steps taken by the Parties in the light of the latest scientific assessments concerning climate change;

(b) Consider the information communicated in accordance with Article 12, paragraph 2, in order to assist the Conference of the Parties in carrying out the reviews required by Article 4 , paragraph 2(d); and

(c) Assist the Conference of the Parties, as appropriate, in the preparation and implementation of its decisions.

Article 11
Financial Mechanism

1. A mechanism for the provision of financial resources on a grant or concessional basis, including for the transfer of technology, is hereby defined. It shall function under the guidance of and be accountable to the Conference of the Parties, which shall decide on its policies, programme priorities and eligibility criteria related to this Convention. Its operation shall be entrusted to one or more existing international entities.

2. The financial mechanism shall have an equitable and balanced representation of all Parties within a transparent system of governance.

3. The Conference of the Parties and the entity or entities entrusted with the operation of the financial mechanism shall agree upon arrangements to give effect to the above paragraphs, which shall include the following:

(a) Modalities to ensure that the funded projects to address climate change are in conformity with the policies, programme priorities and eligibility criteria established by the Conference of the Parties;

(b) Modalities by which a particular funding decision may be reconsidered in light of these policies, programme priorities and eligibility criteria;

(c) Provision by the entity or entities of regular reports to the Conference of the Parties on its funding operations, which is consistent with the requirement for accountability set out in paragraph 1 above; and

(d) Determination in a predictable and identifiable manner of the amount of funding necessary and available for the implementation of this Convention and the conditions under which that amount shall be periodically reviewed.

4. The Conference of the Parties shall make arrangements to implement the above mentioned provisions at its first session, reviewing and taking into account the interim arrangements referred to in Article 21, paragraph 3, and shall decide whether these interim arrangements shall be maintained. Within four years thereafter, the Conference of the Parties shall review the financial mechanism and take appropriate measures.

5. The developed country Parties may also provide and developing country Parties avail themselves of, financial resources related to the implementation of the Convention through bilateral, regional and other multilateral channels.

Article 12
Communication of Information Related to Implementation

1. In accordance with Article 4, paragraph 1, each Party shall communicate to the Conference of the Parties, through the secretariat, the following elements of information:

(a) A national inventory of anthropogenic emissions by sources and removals by sinks of all greenhouse gases not controlled by the Montreal Protocol, to the extent its capacities permit, using comparable methodologies to be promoted and agreed upon by the Conference of the Parties;

(b) A general description of steps taken or envisaged by the Party to implement the Convention; and

(c) Any other information that the Party considers relevant to the achievement of the objective of the Convention and suitable for inclusion in its communication, including, if feasible, material relevant for calculations of global emission trends.

2. Each developed country Party and each other Party included in annex I shall incorporate in its communication the following elements of information:

(a) A detailed description of the policies and measures referred to in subparagraph (a) immediately above will have on anthropogenic emissions by its sources and removals by its sinks of greenhouse gases during the period referred to in Article 4, paragraph 2(a).

3. In addition, each developed country Party and each other developed Party included in annex II shall incorporate details of measures taken in accordance with Article 4, paragraphs 3, 4 and 5.

4. Developing country Parties may, on a voluntary basis, propose projects for financing, including specific technologies, materials, equipment, techniques

or practices that would be needed to implement such projects, along with, if possible, an estimate of all incremental costs, of the reductions of emissions and increments of removals of greenhouse gases, as well as an estimate of the consequent benefits.

5. Each developed country Party and each other Party included in annex I shall make its initial communication within six months of the entry into force of the Convention for that Party. Each Party not so listed shall make its initial communication within three years of the entry into force of the Convention for that Party, or of the availability of financial resources in accordance with Article 4, paragraph 3. Parties that are least developed countries may make their initial communication at their discretion. The frequency of subsequent communications by all Parties shall be determined by the Conference of the Parties, taking into account the differentiated timetable set by this paragraph.

6. Information communicated by Parties under this Article shall be transmitted by the secretariat as soon as possible to the Conference of the Parties and to any subsidiary bodies concerned. If necessary, the procedures for the communication of information may be further considered by the Conference of the Parties.

7. From its first session, the Conference of the Parties shall arrange for the provision to developing country Parties of technical and financial support, on request, in compiling and communicating information under this Article, as well as in identifying the technical and financial needs associated with proposed projects and response measures under Article 4. Such support may be provided by other Parties, by competent internationalorganizations and by the secretariat, as appropriate.

8. Any group of Parties may, subject to guidelines adopted by the Conference of the Parties, and to prior notification to the Conference of the Parties, make a joint communication in fulfilment of their obligations under this Article, provided that such a communication includes information on the fulfilment by each of these Parties of its individual obligations under the Convention.

9. Information received by the secretariat that is designated by a Party as confidential, in accordance with criteria to be established by the Conference of the Parties, shall be aggregated by the secretariat to protect its confidentiality before being made available to any of the bodies involved in the communication and review of information.

10. Subject to paragraph 9 above, and without prejudice to the ability of any Party to make public its communication at any time, the secretariat shall make communications by Parties under this Article publicly available at the time they are submitted to the Conference of the Parties.

Article 13
Resolution of Questions Regarding Implementation

The Conference of the Parties shall, at its first session, consider the establishment of a multilateral consultative process, available to Parties on their request, for the resolution of questions regarding the implementation of the Convention.

Article 14
Settlement of Disputes

1. In the event of a dispute between any two or more Parties concerning the interpretation or application of the Convention, the Parties concerned shall seek a settlement of the dispute through negotiation or any other peaceful means of their own choice.

2. When ratifying, accepting, approving or acceding to the Convention, or at any time thereafter, a Party which is not a regional economic integration organization may declare in a written instrument submitted to the Depositary that, in respect of any dispute concerning the interpretation or application of the Convention, it recognizes as compulsory ipso facto and without special agreement, in relation to any Party accepting the same obligation:

(a) Submission of the dispute to the International Court of Justice, and/or

(b) Arbitration in accordance with procedures to be adopted by the Conference of the Parties as soon as practicable, in an annex on arbitration.

A Party which is a regional economic integration organization may make a declaration with like effect in relation to arbitration in accordance with the procedures referred to in subparagraph (b) above.

3. A declaration made under paragraph 2 above shall remain in force until it expires in accordance with its terms or until three months after written notice of its revocation has been deposited with the Depositary.

4. A new declaration, a notice of revocation or the expiry of a declaration shall not in any way affect proceedings pending before the International Court of Justice or the arbitral tribunal, unless the parties to the dispute otherwise agree.

5. Subject to the operation of paragraph 2 above, if after twelve months following notification by one Party to another that a dispute exists between them, the Parties concerned have not been able to settle their dispute through the means mentioned in paragraph 1 above, the dispute shall be submitted, at the request of any of the parties to the dispute, to conciliation.

6. A conciliation commission shall be created upon the request of one of the parties to the dispute. The commission shall be composed of an equal number of members appointed by each party concerned and a chairman chosen jointly by the members appointed by each party. The commission shall render a recommendatory award, which the parties shall consider in good faith.

7. Additional procedures relating to conciliation shall be adopted by the Conference of the Parties, as soon as practicable, in an annex on conciliation.

8. The provisions of this Article shall apply to any related legal instrument which the Conference of the Parties may adopt, unless the instrument provides otherwise.

Article 15
Amendments to the Convention

1. Any Party may propose amendments to the Convention.

2. Amendments to the Convention shall be adopted at an ordinary session of the Conference of the Parties. The text of any proposed amendment to the Convention shall be communicated to the Parties by the secretariat at least six months before the meeting at which it is proposed for adoption. The secretariat shall also communicate proposed amendments to the signatories to the Convention and, for information, to the Depositary.

3. The Parties shall make every effort to reach agreement on any proposed amendment to the Convention by consensus. If all efforts at consensus have been exhausted, and no agreement reached, the amendment shall as a last resort be adopted by a three-fourths majority vote of the Parties present and voting at the meeting. The adopted amendment shall be communicated by the secretariat to the Depositary, who shall circulate it to all Parties for their acceptance.

4. Instruments of acceptance in respect of an amendment shall be deposited with the Depositary. An amendment adopted in accordance with paragraph 3 above shall enter into force for those Parties having accepted it on the ninetieth day after the date of receipt by the Depositary of an instrument of acceptance by at least three-fourths of the Parties to the Convention.

5. The amendment shall enter into force for any other Party on the ninetieth day after the date on which that Party deposits with theDepositary its instrument of acceptance of the said amendment.

6. For the purposes of this Article, "Parties present and voting" means Parties present and casting an affirmative or negative vote.

Article 16
Adoption and Amendment of Annexes to the Convention

1. Annexes to the Convention shall form an integral part thereof and, unless otherwise expressly provided, a reference to the Convention constitutes at the same time a reference to any annexes thereto. Without prejudice to the provisions of Article 14, paragraphs 2(b) and 7, such annexes shall be restricted to lists, forms and any other material of a descriptive nature that is of a scientific, technical, procedural or administrative character.

2. Annexes to the Convention shall be proposed and adopted in accordance with the procedure set forth in Article 15, paragraphs 2, 3, and 4.

3. An annex that has been adopted in accordance with paragraph 2 above shall enter into force for all Parties to the Convention six months after the date of the communication by the Depositary to such Parties of the adoption of the annex, except for those Parties that have notified the Depositary, in writing, within that period of their non-acceptance of the annex. The annex shall enter into force for Parties which withdraw their notification of non-acceptance on the ninetieth day after the date on which withdrawal of such notification has been received by the Depositary.

4. The proposal, adoption and entry into force of amendments to annexes to the Convention shall be subject to the same procedure as that for the proposal, adoption and entry into force of annexes to the Convention in accordance with paragraphs 2 and 3 above.

5. If the adoption of an annex or an amendment to an annex involves an amendment to the Convention, that annex or amendment to an annex shall not enter into force until such time as the amendment to the Convention enters into force.

Article 17
Protocols

1. The Conference of the Parties may, at any ordinary session, adopt protocols to the Convention.

2. The text of any proposed protocol shall be communicated to the Parties by the secretariat at least six months before such a session.

3. The requirements for the entry into force of any protocol shall be established by that instrument.

4. Only Parties to the Convention may be Parties to a protocol.

5. Decisions under any protocol shall be taken only by the Parties to the protocol concerned.

Article 18
Right to Vote

1. Each Party to the Convention shall have one vote, except as provided for in paragraph 2 below.

2. Regional economic integration organizations, in matters within their competence, shall exercise their right to vote with a number of votes equal to the number of their member States that are Parties to the Convention. Such an organization shall not exercise its right to vote if any of its member States exercises its right, and vice versa.

Article 19
Depositary

The Secretary-General of the United Nations shall be the Depositary of the Convention and of protocols adopted in accordance with Article 17.

Article 20
Signature

This Convention shall be open for signature by States Members of the United Nations or of any of its specialized agencies or that are Parties to the Statute of the International Court of Justice and by regional economic integration organizations at Rio de Janeiro, during the United Nations Conference on Environment and Development, and thereafter at United Nations Headquarters in New York from 20 June 1992 to 19 June 1993.

Article 21
Interim Arrangements

1. The secretariat functions referred to in Article 8 will be carried out on an interim basis by the secretariat established by the General Assembly of the United Nations in its resolution45/212 of 21 December 1990, until the completion of the first session of the Conference of the Parties.

2. The head of the interim secretariat referred to in paragraph 1 above will cooperate closely with the Intergovernmental Panel on Climate Change to ensure that the Panel can respond to the need for objective scientific and technical advice. Other relevant scientific bodies could also be consulted.

3. The Global Environment Facility of the United Nations Development Programme, the United Nations Environment Programme and the International Bank for Reconstruction and Development shall be the international entity entrusted with the operation of the financial mechanism referred to in Article 11 on an interim basis. In this connection, the Global Environment Facility should be appropriately restructured and its membership made universal to enable it to fulfil the requirements of Article 11.

Article 22
Ratification, Acceptance, Approval or Accession

1. The Convention shall be subject to ratification, acceptance, approval or accession by States and by regional economic integrationorganizations. It shall be open for accession from the day after the date on which the Convention is closed for signature. Instruments of ratification, acceptance, approval or accession shall be deposited with the Depositary.

2. Any regional economic integrationorganization which becomes a Party to the Convention without any of its member States being a Party shall be bound by all the obligations under the Convention. In the case of suchorganizations, one or more of whose member States is a Party to the Convention, the organization and its member States shall decide on their respective responsibilities for the performance of their obligations under the Convention. In such cases, the organization and the member States shall not be entitled to exercise rights under the Convention concurrently.

3. In their instruments of ratification, acceptance, approval or accession, regional economic integration organizations shall declare the extent of their competence with respect to the matters governed by the Convention. These organizations shall also inform the Depositary, who shall in turn inform the Parties, of any substantial modification in the extent of their competence.

Article 23
Entry into Force

1. The Convention shall enter into force on the ninetieth day after the date of deposit of the fiftieth instrument of ratification, acceptance, approval or accession.

2. For each State or regional economic integrationorganization that ratifies, accepts or approves the Convention or accedes thereto after the deposit of the fiftieth instrument of ratification, acceptance, approval or accession, the Convention shall enter into force on the ninetieth day after the date of deposit by

such State or regional economic integration organization of its instrument of ratification, acceptance, approval or accession.

3. For the purposes of paragraphs 1 and 2 above, any instrument deposited by a regional economic integration organization shall not be counted as additional to those deposited by States members of the organization.

Article 24
Reservations

No reservations may be made to the Convention.

Article 25
Withdrawal

1. At any time after three years from the date on which the Convention has entered into force for a Party, that Party may withdraw from the Convention by giving written notification to the Depositary.

2. Any such withdrawal shall take effect upon expiry of one year from the date of receipt by the Depositary of the notification of withdrawal, or on such later date as may be specified in the notification of withdrawal.

3. Any Party that withdraws from the Convention shall be considered as also having withdrawn from any protocol to which it is a Party.

Article 26
Authentic Texts

The original of this Convention, of which the Arabic, Chinese, English, French, Russian and Spanish texts are equally authentic, shall be deposited with the Secretary-General of the United Nations.

IN WITNESS WHEREOF the undersigned, being duly authorized to that effect, have signed this Convention.

DONE at New York this ninth day of May one thousand nine hundred and ninety-two.

Annex I

Australia
Austria
Belarus a/
Belgium
Bulgaria a/
Canada
Czechoslovakia a/
Denmark
European Community
Estonia a/
Finland
France
Germany
Greece
Hungary a/
Iceland
Ireland
Italy
Japan
Latvia a/
Lithuania a/
Luxembourg
Netherlands
New Zealand
Norway
Poland a/
Portugal
Romania a/
Russian Federation a/
Spain
Sweden
Switzerland
Turkey
Ukraine a/

United Kingdom of Great
Britain and Northern Ireland
United States of America

a/ Countries that are undergoing the process of transition to a market economy.

Annex II

Australia
Austria
Belgium
Canada
Denmark
European Community
Finland
France
Germany
Greece
Iceland
Ireland
Italy
Japan
Luxembourg
Netherlands
New Zealand
Norway
Portugal
Spain
Sweden
Switzerland
Turkey
United Kingdom of Great
Britain and Northern Ireland
United States of America

Kyoto Protocol to the United Nations Framework Convention on Climate Change

The Parties to this Protocol,

Being Parties to the United Nations Framework Convention on Climate Change, hereinafter referred to as "the Convention",

In pursuit of the ultimate objective of the Convention as stated in its Article 2,

Recalling the provisions of the Convention,

Being guided by Article 3 of the Convention,

Pursuant to the Berlin Mandate adopted by decision 1/CP.1 of the Conference of the Parties to the Convention at its first session,

Have agreed as follows:

Article 1

For the purposes of this Protocol, the definitions contained in Article 1 of the Convention shall apply. In addition:

1. "Conference of the Parties" means the Conference of the Parties to the Convention.

2. "Convention" means the United Nations Framework Convention on Climate Change, adopted in New York on 9 May 1992.

3. "Intergovernmental Panel on Climate Change" means the Intergovernmental Panel on Climate Change established in 1988 jointly by the

World Meteorological Organization and the United Nations Environment Programme.

4. "Montreal Protocol" means the Montreal Protocol on Substances that Deplete the Ozone Layer, adopted in Montreal on 16 September 1987 and as subsequently adjusted and amended.

5. "Parties present and voting" means Parties present and casting an affirmative or negative vote.

6. "Party" means, unless the context otherwise indicates, a Party to this Protocol.

7. "Party included in Annex I" means a Party included in Annex I to the Convention, as may be amended, or a Party which has made a notification under Article 4, paragraph 2(g), of the Convention.

Article 2

1. Each Party included in Annex I in achieving its quantified emission limitation and reduction commitments under Article 3, in order to promote sustainable development, shall:

(a) Implement and/or further elaborate policies and measures in accordance with its national circumstances, such as:

(i) Enhancement of energy efficiency in relevant sectors of the national economy;

(ii) Protection and enhancement of sinks and reservoirs of greenhouse gases not controlled by the Montreal Protocol, taking into account its commitments under relevant international environmental agreements; promotion of sustainable forest management practices, afforestation and reforestation;

(iii) Promotion of sustainable forms of agriculture in light of climate change considerations;

(iv) Promotion, research, development and increased use of new and renewable forms of energy, of carbon dioxide sequestration technologies and of advanced and innovative environmentally sound technologies;

(v) Progressive reduction or phasing out of market imperfections, fiscal incentives, tax and duty exemptions and subsidies in all greenhouse gas emitting sectors that run counter to the objective of the Convention and apply market instruments;

(vi) Encouragement of appropriate reforms in relevant sectors aimed at promoting policies and measures which limit or reduce emissions of greenhouse gases not controlled by the Montreal Protocol;

(vii) Measures to limit and/or reduce emissions of greenhouse gases not controlled by the Montreal Protocol in the transport sector;

(viii) Limitation and/or reduction of methane through recovery and use in waste management, as well as in the production, transport and distribution of energy;

(b) Cooperate with other such Parties to enhance the individual and combined effectiveness of their policies and measures adopted under this Article, pursuant to Article 4, paragraph 2(e)(i), of the Convention. To this end, these Parties shall take steps to share their experience and exchange information on such policies and measures, including developing ways of improving their comparability, transparency and effectiveness. The Conference of the Parties serving as the meeting of the Parties to this Protocol shall, at its first session or as soon as practicable thereafter, consider ways to facilitate such cooperation, taking into account all relevant information.

2. The Parties included in Annex I shall pursue limitation or reduction of emissions of greenhouse gases not controlled by the Montreal Protocol from aviation and marine bunker fuels, working through the International Civil Aviation Organization and the International Maritime Organization, respectively.

3. The Parties included in Annex I shall strive to implement policies and measures under this Article in such a way as to minimize adverse effects, including the adverse effects of climate change, effects on international trade, and social, environmental and economic impacts on other Parties, especially developing country Parties and in particular those identified in Article 4, paragraphs 8 and 9 of the Convention, taking into account Article 3 of the Convention. The Conference of the Parties serving as the meeting of the Parties to this Protocol may take further action, as appropriate, to promote the implementation of the provisions of this paragraph.

4. The Conference of the Parties serving as the meeting of the Parties to this Protocol, if it decides that it would be beneficial tocoordinate any of the policies and measures in paragraph 1(a) above, taking into account different national circumstances and potential effects, shall consider ways and means to elaborate the coordination of such policies and measures.

Article 3

1. The Parties included in Annex I shall, individually or jointly, ensure that their aggregate anthropogenic carbon dioxide equivalent emissions of the greenhouse gases listed in Annex A do not exceed their assigned amounts, calculated pursuant to their quantified emission limitation and reduction commitments inscribed in Annex B and in accordance with the provisions of this Article, with a view to reducing their overall emissions of such gases by at least 5 per cent below 1990 levels in the commitment period 2008 to 2012.

2. Each Party included in Annex I shall, by 2005, have made demonstrable progress in achieving its commitments under this Protocol.

3. The net changes in greenhouse gas emissions from sources and removals by sinks resulting from direct human-induced land use change and forestry activities, limited to afforestation, reforestation, and deforestation since 1990, measured as verifiable changes in stocks in each commitment period shall be used to meet the commitments in this Article of each Party included in Annex I. The greenhouse gas emissions from sources and removals by sinks associated with those activities shall be reported in a transparent and verifiable manner and reviewed in accordance with Articles 7 and 8.

4. Prior to the first session of the Conference of the Parties serving as the meeting of the Parties to this Protocol, each Party included in Annex I shall provide for consideration by the Subsidiary Body for Scientific and Technological Advice data to establish its level of carbon stocks in 1990 and to enable an estimate to be made of its changes in carbon stocks in subsequent years. The Conference of the Parties serving as the meeting of the Parties to this Protocol shall, at its first session or as soon as practicable thereafter, decide upon modalities, rules and guidelines as to how and which additional human-induced activities related to changes in greenhouse gas emissions and removals in the agricultural soil and land use change and forestry categories, shall be added to, or subtracted from, the assigned amount for Parties included in Annex I, taking into account uncertainties, transparency in reporting, verifiability, the methodological work of the Intergovernmental Panel on Climate Change, the advice provided by the Subsidiary Body for Scientific and Technological Advice in accordance with Article 5 and the decisions of the Conference of the Parties. Such a decision shall apply in the second and subsequent commitment periods. A Party may choose to apply such a decision on these additional human-induced activities for its first commitment period, provided that these activities have taken place since 1990.

5. The Parties included in Annex I undergoing the process of transition to a market economy whose base year or period was established pursuant to decision 9/CP.2 of the Conference of the Parties at its second session, shall use that base year or period for the implementation of their commitments under this Article

Any other Party included in Annex I undergoing the process of transition to a market economy which has not yet submitted its first national communication under Article 12 of the Convention may also notify the Conference of the Parties serving as the meeting of the Parties to this Protocol that it intends to use a historical base year or period other than 1990 for the implementation of its commitments under this Article. The Conference of the Parties serving as the meeting of the Parties to this Protocol shall decide on the acceptance of such notification.

6. Taking into account Article 4, paragraph 6, of the Convention, in the implementation of their commitments under this Protocol other than those in this Article, a certain degree of flexibility shall be allowed by the Conference of the Parties serving as the meeting of the Parties to this Protocol to the Parties included in Annex I undergoing the process of transition to a market economy.

7. In the first quantified emission limitation and reduction commitment period, from 2008 to 2012, the assigned amount for each Party included in Annex I shall be equal to the percentage inscribed for it in Annex B of its aggregate anthropogenic carbon dioxide equivalent emissions of the greenhouse gases listed in Annex A in 1990, or the base year or period determined in accordance with paragraph 5 above, multiplied by five. Those Parties included in Annex I for whom land use change and forestry constituted a net source of greenhouse gas emissions in 1990 shall include in their 1990 emissions base year or period the aggregate anthropogenic carbon dioxide equivalent emissions minus removals in 1990 from land use change for the purposes of calculating their assigned amount.

8. Any Party included in Annex I may use 1995 as its base year for hydrofluorocarbons, perfluorocarbons and sulphur hexafluoride, for the purposes of the calculation referred to in paragraph 7 above.

9. Commitments for subsequent periods for Parties included in Annex I shall be established in amendments to Annex B to this Protocol, which shall be adopted in accordance with the provisions of Article 20, paragraph 7. The Conference of the Parties serving as the meeting of the Parties to this Protocol

shall initiate the consideration of such commitments at least seven years before the end of the first commitment period mentioned in paragraph 7 above.

10. Any emission reduction units, or any part of an assigned amount, which a Party acquires from another Party in accordance with the provisions of Article 6 and of Article 16 bis shall be added to the assigned amount for that Party.

11. Any emission reduction units, or any part of an assigned amount, which a Party transfers to another Party in accordance with the provisions of Article 6 and of Article 16 bis shall be subtracted from the assigned amount for that Party.

12. Any certified emission reductions which a Party acquires from another Party in accordance with the provisions of Article 12 shall be added to the assigned amount for that Party.

13. If the emissions of a Party included in Annex I during a commitment period are less than its assigned amount under this Article, this difference shall, on request of that Party, be added to the assigned amount for that Party for subsequent commitment periods.

14. Each Party included in Annex I shall strive to implement the commitments mentioned in paragraph 1 above in such a way as to minimize adverse social, environmental and economic impacts on developing country Parties, particularly those identified in Article 4, paragraphs 8 and 9, of the Convention. In line with relevant decisions of the Conference of the Parties on the implementation of those paragraphs, the Conference of the Parties serving as the meeting of the Parties to this Protocol shall, at its first session, consider what actions are necessary to minimize the adverse effects of climate change and/or the impacts of response measures on Parties referred to in those paragraphs. Among the issues to be considered shall be the establishment of funding, insurance and transfer of technology.

Article 4

1. Any Parties included in Annex I that have agreed to jointly fulfil their commitments under Article 3 shall be deemed to have met those commitments provided that their total combined aggregate anthropogenic carbon dioxide equivalent emissions of the greenhouse gases listed in Annex A do not exceed their assigned amounts calculated pursuant to their quantified emission limitation and reduction commitments inscribed in Annex B and in accordance with the provisions of Article 3. The respective emission level allocated to each of the Parties to the agreement shall be set out in that agreement.

2. The Parties to any such agreement shall notify the secretariat of the terms of the agreement on the date of deposit of their instruments of ratification, acceptance, approval or accession. The secretariat shall in turn inform the Parties and signatories to the Convention of the terms of the agreement.

3. The agreement shall remain in operation for the duration of the commitment period specified in Article 3, paragraph 7.

4. If Parties acting jointly do so in the framework of, and together with, a regional economic integrationorganization, any alteration in the composition of the organization after adoption of this Protocol shall not affect existing commitments under this Protocol. Any alteration in the composition of the organization shall only apply for the purposes of those commitments under Article 3 that are adopted subsequent to that revision.

5. In the event of failure by the Parties to such an agreement to achieve their total combined level of emission reductions, each Party to such an agreement shall be responsible for its own level of emissions set out in the agreement.

6. If Parties acting jointly do so in the framework of, and together with, a regional economic integration organization which is itself a Party to this Protocol, each member State of that regional economic integrationorganization individually, and together with the regional economic integrationorganization acting in accordance with Article 23, shall, in the event of failure to achieve the

total combined level of emission reductions, be responsible for its level of emissions as notified in accordance with this Article.

Article 5

1. Each Party included in Annex I shall have in place, no later than one year prior to the start of the first commitment period, a national system for the estimation of anthropogenic emissions by sources and removals by sinks of all greenhouse gases not controlled by the Montreal Protocol. Guidelines for such national systems, which shall incorporate the methodologies specified in paragraph 2 below, shall be decided upon by the Conference of the Parties serving as the meeting of the Parties to this Protocol at its first session.

2. Methodologies for estimating anthropogenic emissions by sources and removals by sinks of all greenhouse gases not controlled by the Montreal Protocol shall be those accepted by the Intergovernmental Panel on Climate Change and agreed upon by the Conference of the Parties at its third session. Where such methodologies are not used, appropriate adjustments shall be applied according to methodologies agreed upon by the Conference of the Parties serving as the meeting of the Parties to this Protocol at its first session. Based on the work of, inter alia, the Intergovernmental Panel on Climate Change and advice provided by the Subsidiary Body for Scientific and Technological Advice, the Conference of the Parties serving as the meeting of the Parties to this Protocol shall regularly review and, as appropriate, revise such methodologies and adjustments, taking fully into account any relevant decisions by the Conference of the Parties. Any revision to methodologies or adjustments shall be used only for the purposes of ascertaining compliance with commitments under Article 3 in respect of any commitment period adopted subsequent to that revision.

3. The global warming potentials used to calculate the carbon dioxide equivalence of anthropogenic emissions by sources and removals by sinks of greenhouse gases not controlled by the Montreal Protocol listed in Annex A shall be those accepted by the Intergovernmental Panel on Climate Change and agreed upon by the Conference of the Parties at its third session. Based on the

work of, inter alia, the Intergovernmental Panel on Climate Change and advice provided by the Subsidiary Body for Scientific and Technological Advice, the Conference of the Parties serving as the meeting of the Parties to this Protocol shall regularly review and, as appropriate, revise the global warming potential of each such greenhouse gas, taking fully into account any relevant decisions by the Conference of the Parties. Any revision to a global warming potential shall apply only to those commitments under Article 3 in respect of any commitment period adopted subsequent to that revision.

Article 6

1. For the purpose of meeting its commitments under Article 3, any Party included in Annex I may transfer to, or acquire from, any other such Party emission reduction units resulting from projects aimed at reducing anthropogenic emissions by sources or enhancing anthropogenic removals by sinks of greenhouse gases in any sector of the economy, provided that:

(a) Any such project has the approval of the Parties involved;

(b) Any such project provides a reduction in emissions by sources, or an enhancement of removals by sinks, that is additional to any that would otherwise occur;

(c) It does not acquire any emission reduction units if it is not in compliance with its obligations under Articles 5 and 7; and

(d) The acquisition of emission reduction units shall be supplemental to domestic actions for the purposes of meeting commitments under Article 3.

2. The Conference of the Parties serving as the meeting of the Parties to this Protocol may, at its first session or as soon as practicable thereafter, further elaborate guidelines for the implementation of this Article, including for verification and reporting.

3. A Party included in Annex I may authorize legal entities to participate, under its responsibility, in actions leading to the generation, transfer or acquisition under this Article of emission reduction units.

4. If a question of implementation by a Party included in Annex I of the requirements referred to in this paragraph is identified in accordance with the relevant provisions of Article 8, transfers and acquisitions of emission reduction units may continue to be made after the question has been identified, provided that any such units may not be used by a Party to meet its commitments under Article 3 until any issue of compliance is resolved.

Article 7

1. Each Party included in Annex I shall incorporate in its annual inventory of anthropogenic emissions by sources and removals by sinks of greenhouse gases not controlled by the Montreal Protocol, submitted in accordance with the relevant decisions of the Conference of the Parties, the necessary supplementary information for the purposes of ensuring compliance with Article 3, to be determined in accordance with paragraph 4 below.

2. Each Party included in Annex I shall incorporate in its national communication, submitted under Article 12 of the Convention, the supplementary information necessary to demonstrate compliance with its commitments under this Protocol, to be determined in accordance with paragraph 4 below.

3. Each Party included in Annex I shall submit the information required under paragraph 1 above annually, beginning with the first inventory due under the Convention for the first year of the commitment period after this Protocol has entered into force for it. Each such Party shall submit the information required under paragraph 2 above as part of the first national communication due under the Convention after this Protocol has entered into force for it and after the adoption of guidelines as provided for in paragraph 4 below. The frequency of subsequent submission of information required under this Article shall be determined by the Conference of the Parties serving as the meeting of

the Parties to this Protocol, taking into account any timetable for the submission of national communications decided upon by the Conference of the Parties.

4. The Conference of the Parties serving as the meeting of the Parties to this Protocol shall adopt at its first session, and review periodically thereafter, guidelines for the preparation of the information required under this Article, taking into account guidelines for the preparation of national communications by Parties included in Annex I adopted by the Conference of the Parties. The Conference of the Parties serving as the meeting of the Parties to this Protocol shall also, prior to the first commitment period, decide upon modalities for the accounting of assigned amounts.

Article 8

1. The information submitted under Article 7 by each Party included in Annex I shall be reviewed by expert review teams pursuant to the relevant decisions of the Conference of the Parties and in accordance with guidelines adopted for this purpose by the Conference of the Parties serving as the meeting of the Parties to this Protocol under paragraph 4 below. The information submitted under Article 7, paragraph 1, by each Party included in Annex I shall be reviewed as part of the annual compilation and accounting of emissions inventories and assigned amounts. Additionally, the information submitted under Article 7, paragraph 2, by each Party included in Annex I shall be reviewed as part of the review of communications.

2. Expert review teams shall be coordinated by the secretariat and shall be composed of experts selected from those nominated by Parties to the Convention and, as appropriate, by intergovernmental organizations, in accordance with guidance provided for this purpose by the Conference of the Parties.

3. The review process shall provide a thorough and comprehensive technical assessment of all aspects of the implementation by a Party of this Protocol. The expert review teams shall prepare a report to the Conference of the Parties serving as the meeting of the Parties to this Protocol, assessing the

implementation of the commitments of the Party and identifying any potential problems in, and factors influencing, the fulfilment of commitments. Such reports shall be circulated by the secretariat to all Parties to the Convention. The secretariat shall list those questions of implementation indicated in such reports for further consideration by the Conference of the Parties serving as the meeting of the Parties to this Protocol.

4. The Conference of the Parties serving as the meeting of the Parties to this Protocol shall adopt at its first session, and review periodically thereafter, guidelines for the review of implementation by expert review teams taking into account the relevant decisions of the Conference of the Parties.

5. The Conference of the Parties serving as the meeting of the Parties to this Protocol shall, with the assistance of the Subsidiary Body for Implementation and, as appropriate, the Subsidiary Body for Scientific and Technological Advice, consider:

(a) The information submitted by the Parties under Article 7 and the reports of the expert reviews thereon conducted under this Article; and

(b) Those questions of implementation listed by the secretariat under paragraph 3 above, as well as any questions raised by Parties.

6. Pursuant to its consideration of the information referred to in paragraph 5 above, the Conference of the Parties serving as the meeting of the Parties to this Protocol shall take decisions on any matter required for the implementation of this Protocol.

Article 9

1. The Conference of the Parties serving as the meeting of the Parties to this Protocol shall periodically review this Protocol in the light of the best available scientific information and assessments on climate change and its imp well as relevant technical, social and economic information. Such rev be coordinated with pertinent reviews under the Convention, in particular those

required by Article 4, paragraph 2(d), and Article 7, paragraph 2(a), of the Convention. Based on these reviews, the Conference of the Parties serving as the meeting of the Parties to this Protocol shall take appropriate action.

2. The first review shall take place at the second session of the Conference of the Parties serving as the meeting of the Parties to this Protocol. Further reviews shall take place at regular intervals and in a timely manner.

Article 10

All Parties, taking into account their common but differentiated responsibilities and their specific national and regional development priorities, objectives and circumstances, without introducing any new commitments for Parties not included in Annex I, but reaffirming existing commitments in Article 4, paragraph 1, of the Convention, and continuing to advance the implementation of these commitments in order to achieve sustainable development, taking into account Article 4, paragraphs 3, 5 and 7, of the Convention, shall:

(a) Formulate, where relevant and to the extent possible, cost-effective national, and where appropriate regional programmes to improve the quality of local emission factors, activity data and/or models which reflect the socio-economic conditions of each Party for the preparation and periodic updating of national inventories of anthropogenic emissions by sources and removals by sinks of all greenhouse gases not controlled by the Montreal Protocol, using comparable methodologies to be agreed upon by the Conference of the Parties, and consistent with the guidelines for national communications adopted by the Conference of the Parties;

(b) Formulate, implement, publish and regularly update national and, where appropriate, regional programmes containing measures to mitigate climate change and measures to facilitate adequate adaptation to climate change:

(i) Such programmes would, inter alia, concern the energy, transport and industry sectors as well as agriculture, forestry and waste management. Furthermore, adaptation technologies and methods for improving spatial planning would improve adaptation to climate change; and

(ii) Parties included in Annex I shall submit information on action under this Protocol, including national programmes, according to the guidelines laid down in Article 8; and other Parties shall seek to include in their national communications, as appropriate, information on programmes which contain measures that the Party believes contribute to addressing climate change and its adverse impacts, including the abatement of increase in greenhouse gas emissions, and enhancement of and removals by sinks, capacity building and adaptation measures.

(c) Cooperate in the promotion of effective modalities for the development, application and diffusion of, and take all practicable steps to promote, facilitate and finance, as appropriate, the transfer of, or access to, environmentally sound technologies, know-how, practices and processes pertinent to climate change, in particular to developing countries, including the formulation of policies and programmes for the effective transfer of environmentally sound technologies that are publicly owned or in the public domain and the creation of an enabling environment for the private sector, to promote and enhance access to, and transfer of, environmentally sound technologies;

(d) Cooperate in scientific and technical research and promote the maintenance and the development of systematic observation systems and development of data archives to reduce uncertainties related to the climate system, the adverse impacts of climate change and the economic and social consequences of various response strategies, and promote the development and strengthening of endogenous capacities and capabilities to participate in international and intergovernmental efforts, programmes and networks on research and systematic observation, taking into account Article 5 of the Convention;

(e) Cooperate in and promote at the international level, and, where appropriate, using existing bodies, the development and implementation of education and training programmes, including the strengthening of national capacity building, in particular human and institutional capacities and the exchange or secondment of personnel to train experts in this field, in particular for developing countries, and facilitate at the national level public awareness and public access to information on climate change. Suitable modalities should be developed to implement these activities through the relevant bodies of the Convention taking into account Article 6 of the Convention;

(f) Include in their national communications information on programmes and activities undertaken pursuant to this Article in accordance with relevant decisions of the Conference of the Parties; and

(g) Give full consideration, in implementing the commitments in this Article, to Article 4, paragraph 8, of the Convention.

Article 11

1. In the implementation of Article 10, Parties shall take into account the provisions of Article 4, paragraphs 4, 5, 7, 8 and 9 of the Convention.

2. In the context of the implementation of Article 4, paragraph 1, of the Convention, in accordance with the provisions of Article 4, paragraph 3, and Article 11 of the Convention, and through the operating entity or entities of the financial mechanism of the Convention, the developed country Parties and other developed Parties included in Annex II to the Convention shall:

(a) Provide new and additional financial resources to meet the agreed full costs incurred by developing country Parties in advancing the implementation of existing commitments under Article 4, paragraph 1(a), of the Convention that are covered in Article 10, subparagraph (a); and

(b) Also provide such financial resources, including for the transfer of technology, needed by the developing country Parties to meet the agreed full incremental costs of advancing the implementation of existing commitments in Article 4, paragraph 1, of the Convention that are covered by Article 10 and that are agreed between a developing country Party and the international entity or entities referred to in Article 11 of the Convention, in accordance with that Article.

The implementation of these existing commitments shall take into account the need for adequacy and predictability in the flow of funds and the importance of appropriate burden sharing among developed country Parties. The guidance to the financial mechanism of the Convention in relevant decisions of the Conference of the Parties, including those agreed before the adoption of this Protocol, shall apply mutatis mutandis to the provisions of this paragraph.

3. The developed country Parties and other developed Parties in Annex II to the Convention may also provide, and developing country Parties avail themselves of, financial resources for the implementation of Article 10, through bilateral, regional and other multilateral channels.

Article 12

1. A clean development mechanism is hereby defined.

2. The purpose of the clean development mechanism shall be to assist Parties not included in Annex I in achieving sustainable development and in contributing to the ultimate objective of the Convention, and to assist Parties included in Annex I in achieving compliance with their quantified emission limitation and reduction commitments under Article 3.

3. Under the clean development mechanism:

(a) Parties not included in Annex I will benefit from project activities resulting in certified emission reductions; and

(b) Parties included in Annex I may use the certified emission reductions accruing from such project activities to contribute to compliance with part of their quantified emission limitation and reduction commitments under Article 3, as determined by the Conference of the Parties serving as the meeting of the Parties to this Protocol.

4. The clean development mechanism shall be subject to the authority and guidance of the Conference of the Parties serving as the meeting of the Parties to this Protocol and be supervised by an executive board of the clean development mechanism.

5. Emission reductions resulting from each project activity shall be certified by operational entities to be designated by the Conference of the Parties serving as the meeting of the Parties to this Protocol, on the basis of:

(a) Voluntary participation approved by each Party involved;

(b) Real, measurable, and long-term benefits related to the mitigation of climate change; and

(c) Reductions in emissions that are additional to any that would occur in the absence of the certified project activity.

6. The clean development mechanism shall assist in arranging funding of certified project activities as necessary.

7. The Conference of the Parties serving as the meeting of the Parties to this Protocol shall, at its first session, elaborate modalities and procedures with the objective of ensuring transparency, efficiency and accountability through independent auditing and verification of project activities.

8. The Conference of the Parties serving as the meeting of the Parties to this Protocol shall ensure that a share of the proceeds from certified project activities

is used to cover administrative expenses as well as to assist developing country Parties that are particularly vulnerable to the adverse effects of climate change to meet the costs of adaptation.

9. Participation under the clean development mechanism, including in activities mentioned in paragraph 3(a) above and acquisition of certified emission reductions, may involve private and/or public entities, and is to be subject to whatever guidance may be provided by the executive board of the clean development mechanism.

10. Certified emission reductions obtained during the period from the year 2000 up to the beginning of the first commitment period can be used to assist in achieving compliance in the first commitment period.

Article 13

1. The Conference of the Parties, the supreme body of the Convention, shall serve as the meeting of the Parties to this Protocol.

2. Parties to the Convention that are not Parties to this Protocol may participate as observers in the proceedings of any session of the Conference of the Parties serving as the meeting of the Parties to this Protocol. When the Conference of the Parties serves as the meeting of the Parties to this Protocol, decisions under this Protocol shall be taken only by those that are Parties to it.

3. When the Conference of the Parties serves as the meeting of the Parties to this Protocol, any member of the Bureau of the Conference of the Parties representing a Party to the Convention but, at that time, not a Party to this Protocol, shall be substituted by an additional member to be elected by and from amongst the Parties to this Protocol.

4. The Conference of the Parties serving as the meeting of the Parties to this Protocol shall keep under regular review the implementation of this Protocol and shall make, within its mandate, the decisions necessary to promote its effective

implementation. It shall perform the functions assigned to it by this Protocol and shall:

(a) Assess, on the basis of all information made available to it in accordance with the provisions of this Protocol, the implementation of this Protocol by the Parties, the overall effects of the measures taken pursuant to this Protocol, in particular environmental, economic and social effects as well as their cumulative impacts and the extent to which progress towards the objective of the Convention is being achieved;

(b) Periodically examine the obligations of the Parties under this Protocol, giving due consideration to any reviews required by Article 4, paragraph 2(d), and Article 7, paragraph 2, of the Convention, in the light of the objective of the Convention, the experience gained in its implementation and the evolution of scientific and technological knowledge, and in this respect consider and adopt regular reports on the implementation of this Protocol;

(c) Promote and facilitate the exchange of information on measures adopted by the Parties to address climate change and its effects, taking into account the differing circumstances, responsibilities and capabilities of the Parties and their respective commitments under this Protocol;

(d) Facilitate, at the request of two or more Parties, the coordination of measures adopted by them to address climate change and its effects, taking into account the differing circumstances, responsibilities and capabilities of the Parties and their respective commitments under this Protocol;

(e) Promote and guide, in accordance with the objective of the Convention and the provisions of this Protocol, and taking fully into account the relevant decisions by the Conference of the Parties, the development and periodic refinement of comparable methodologies for the effective implementation of this Protocol, to be agreed on by the Conference of the Parties serving as the meeting of the Parties to this Protocol;

(f) Make recommendations on any matters necessary for the implementation of this Protocol;

(g) Seek to mobilize additional financial resources in accordance with Article 11, paragraph 2;

(h) Establish such subsidiary bodies as are deemed necessary for the implementation of this Protocol;

(i) Seek and utilize, where appropriate, the services and cooperation of, and information provided by, competent international organizations and intergovernmental and non-governmental bodies; and

(j) Exercise such other functions as may be required for the implementation of this Protocol, and consider any assignment resulting from a decision by the Conference of the Parties.

5. The rules of procedure of the Conference of the Parties and financial procedures of the Convention shall be applied mutatis mutandis under this Protocol, except as may be otherwise decided by consensus by the Conference of the Parties serving as the meeting of the Parties to this Protocol.

6. The first session of the Conference of the Parties serving as the meeting of the Parties to this Protocol shall be convened by the secretariat in conjunction with the first session of the Conference of the Parties that is scheduled after the date of the entry into force of this Protocol. Subsequent ordinary sessions of the Conference of the Parties serving as the meeting of the Parties to this Protocol shall be held every year and in conjunction with ordinary sessions of the Conference of the Parties unless otherwise decided by the Conference of the Parties serving as the meeting of the Parties to this Protocol.

7. Extraordinary sessions of the Conference of the Parties serving as the meeting of the Parties to this Protocol shall be held at such other times as may be deemed necessary by the Conference of the Parties serving as the meeting of the Parties to this Protocol, or at the written request of any Party, provided that, within six months of the request being communicated to the Parties by the secretariat, it is supported by at least one third of the Parties.

8. The United Nations, its specialized agencies and the International Atomic Energy Agency, as well as any State member thereof or observers thereto not party to the Convention, may be represented at sessions of the Conference of the Parties serving as the meeting of the Parties to this Protocol as observers. Any body or agency, whether national or international, governmental or non-governmental, which is qualified in matters covered by this Protocol and which has informed the secretariat of its wish to be represented at a session of the Conference of the Parties serving as the meeting of the Parties to this Protocol as an observer, may be so admitted unless at least one third of the Parties present object. The admission and participation of observers shall be subject to the rules of procedure, as referred to in paragraph 5 above.

Article 14

1. The secretariat established by Article 8 of the Convention shall serve as the secretariat of this Protocol.

2. Article 8, paragraph 2, of the Convention on the functions of the secretariat, and Article 8, paragraph 3, of the Convention on arrangements made for the functioning of the secretariat, shall apply mutatis mutandis to this Protocol. The secretariat shall, in addition, exercise the functions assigned to it under this Protocol.

Article 15

1. The Subsidiary Body for Scientific and Technological Advice and the Subsidiary Body for Implementation established by Articles 9 and 10 of the Convention shall serve as, respectively, the Subsidiary Body for Scientific and Technological Advice and the Subsidiary Body for Implementation of this Protocol. The provisions relating to the functioning of these two bodies under the Convention shall apply mutatis mutandis to this Protocol. Sessions of the meetings of the Subsidiary Body for Scientific and Technological Advice and the Subsidiary Body for Implementation of this Protocol shall be held in conjunction with the meetings of, respectively, the Subsidiary Body for

Scientific and Technological Advice and the Subsidiary Body for Implementation of the Convention.

2. Parties to the Convention that are not Parties to this Protocol may participate as observers in the proceedings of any session of the subsidiary bodies. When the subsidiary bodies serve as the subsidiary bodies of this Protocol, decisions under this Protocol shall be taken only by the Parties to this Protocol.

3. When the subsidiary bodies established by Articles 9 and 10 of the Convention exercise their functions with regard to matters concerning this Protocol, any member of the Bureaux of those subsidiary bodies representing a Party to the Convention but, at that time, not a party to this Protocol, shall be substituted by an additional member to be elected by and from amongst the Parties to this Protocol.

Article 16

The Conference of the Parties serving as the meeting of the Parties to this Protocol shall, as soon as practicable, consider the application to this Protocol of, and modify as appropriate, the multilateral consultative process referred to in Article 13 of the Convention, in the light of any relevant decisions that may be taken by the Conference of the Parties. Any multilateral consultative process that may be applied to this Protocol shall operate without prejudice to the procedures and mechanisms established in accordance with Article 17.

Article 16bis

The Conference of the Parties shall define the relevant principles, modalities, rules and guidelines, in particular for verification, reporting and accountability for emissions trading. The Parties included in Annex B may participate in emissions trading for the purposes of fulfilling their commitments under Article 3 of this Protocol. Any such trading shall be supplemental to domestic actions for the purpose of meeting quantified emission limitation and reduction commitments under that Article.

Article 17

The Conference of the Parties serving as the meeting of the Parties to this Protocol shall, at its first session, approve appropriate and effective procedures and mechanisms to determine and to address cases of non-compliance with the provisions of this Protocol, including through the development of an indicative list of consequences, taking into account the cause, type, degree and frequency of non-compliance. Any procedures and mechanisms under this Article entailing binding consequences shall be adopted by means of an amendment to this Protocol.

Article 18

The provisions of Article 14 of the Convention on settlement of disputes shall apply mutatis mutandis to this Protocol.

Article 19

1. Any Party may propose amendments to this Protocol.

2. Amendments to this Protocol shall be adopted at an ordinary session of the Conference of the Parties serving as the meeting of the Parties to this Protocol. The text of any proposed amendment to this Protocol shall be communicated to the Parties by the secretariat at least six months before the meeting at which it is proposed for adoption. The secretariat shall also communicate the text of any proposed amendments to the Parties and signatories to the Convention and, for information, to the Depositary.

3. The Parties shall make every effort to reach agreement on any proposed amendment to this Protocol by consensus. If all efforts at consensus have been

exhausted, and no agreement reached, the amendment shall as a last resort be adopted by a three-fourths majority vote of the Parties present and voting at the meeting. The adopted amendment shall be communicated by the secretariat to the Depositary, who shall circulate it to all Parties for their acceptance.

4. Instruments of acceptance in respect of an amendment shall be deposited with the Depositary. An amendment adopted in accordance with paragraph 3 above shall enter into force for those Parties having accepted it on the ninetieth day after the date of receipt by the Depositary of an instrument of acceptance by at least three fourths of the Parties to this Protocol.

5. The amendment shall enter into force for any other Party on the ninetieth day after the date on which that Party deposits with theDepositary its instrument of acceptance of the said amendment.

Article 20

1. Annexes to this Protocol shall form an integral part thereof and, unless otherwise expressly provided, a reference to this Protocol constitutes at the same time a reference to any annexes thereto. Any annexes adopted after the entry into force of this Protocol shall be restricted to lists, forms and any other material of a descriptive nature that is of a scientific, technical, procedural or administrative character.

2. Any Party may make proposals for an annex to this Protocol and may propose amendments to annexes to this Protocol.

3. Annexes to this Protocol and amendments to annexes to this Protocol shall be adopted at an ordinary session of the Conference of the Parties serving as the meeting of the Parties to this Protocol. The text of any proposed annex or amendment to an annex shall be communicated to the Parties by the secretariat at least six months before the meeting at which it is proposed for adoption. The secretariat shall also communicate the text of any proposed annex or amendment to an annex to the Parties and signatories to the Convention and, for information, to the Depositary.

4. The Parties shall make every effort to reach agreement on any proposed annex or amendment to an annex by consensus. If all efforts at consensus have been exhausted, and no agreement reached, the annex or amendment to an annex shall as a last resort be adopted by a three-fourths majority vote of the Parties present and voting at the meeting. The adopted annex or amendment to an annex shall be communicated by the secretariat to theDepositary, who shall circulate it to all Parties for their acceptance.

5. An annex, other than Annex A or B, that has been adopted or amended in accordance with paragraphs 3 and 4 above shall enter into force for all Parties to this Protocol six months after the date of the communication by theDepositary to such Parties of the adoption or amendment of the annex, except for those Parties that have notified the Depositary in writing within that period of their non-acceptance of the annex or amendment to the annex. The annex or amendment to an annex shall enter into force for Parties which withdraw their notification of non-acceptance on the ninetieth day after the date on which withdrawal of such notification has been received by the Depositary.

6. If the adoption of an annex or an amendment to an annex involves an amendment to this Protocol, that annex or amendment to an annex shall not enter into force until such time as the amendment to this Protocol enters into force.

7. Amendments to Annexes A and B to this Protocol shall be adopted and enter into force in accordance with the procedure set out in Article 19, provided that any amendments to Annex B shall be adopted only with the written consent of the Party concerned.

Article 21

1. Each Party shall have one vote, except as provided for in paragraph 2 below.

2. Regional economic integration organizations, in matters within their competence, shall exercise their right to vote with a number of votes equal to the number of their member States which are Parties to this Protocol. Such an organization shall not exercise its right to vote if any of its member States exercises its right, and vice versa.

Article 22

The Secretary-General of the United Nations shall be the Depositary of this Protocol.

Article 23

1. This Protocol shall be open for signature and subject to ratification, acceptance or approval by States and regional economic integration organizations which are Parties to the Convention. It shall be open for signature at United Nations Headquarters in New York from 16 March 1998 to 15 March 1999. This Protocol shall be open for accession from the day after the date on which it is closed for signature. Instruments of ratification, acceptance, approval or accession shall be deposited with the Depositary.

2. Any regional economic integration organization which becomes a Party to this Protocol without any of its member States being a Party shall be bound by all the obligations under this Protocol. In the case of suchorganizations, one or more of whose member States is a Party to this Protocol, theorganization and its member States shall decide on their respective responsibilities for the performance of their obligations under this Protocol. In such cases, the organization and the member States shall not be entitled to exercise rights under this Protocol concurrently.

3. In their instruments of ratification, acceptance, approval or accession, regional economic integration organizations shall declare the extent of their competence with respect to the matters governed by this Protocol. These organizations shall also inform the Depositary, who shall in turn

inform the Parties, of any substantial modification in the extent of their competence.

Article 24

1. This Protocol shall enter into force on the ninetieth day after the date on which not less than 55 Parties to the Convention, incorporating Parties included in Annex I which accounted in total for at least 55 per cent of the total carbon dioxide emissions for 1990 of the Parties included in Annex I, have deposited their instruments of ratification, acceptance, approval or accession.

2. For the purposes of this Article, "the total carbon dioxide emissions for 1990 of the Parties included in Annex I" means the amount communicated on or before the date of adoption of this Protocol by the Parties included in Annex I in their first national communications submitted in accordance with Article 12 of the Convention.

3. For each State or regional economic integration organization that ratifies, accepts or approves this Protocol or accedes thereto after the conditions set out in paragraph 1 above for the entry into force have been fulfilled, this Protocol shall enter into force on the ninetieth day following the date of deposit of its instrument of ratification, acceptance, approval or accession.

4. For the purposes of this Article, any instrument deposited by a regional economic integration organization shall not be counted as additional to those deposited by States members of the organization.

Article 25

No reservations may be made to this Protocol.

Article 26

1. At any time after three years from the date on which this Protocol has entered into force for a Party, that Party may withdraw from this Protocol by giving written notification to the Depositary.

2. Any such withdrawal shall take effect upon expiry of one year from the date of receipt by the Depositary of the notification of withdrawal, or on such later date as may be specified in the notification of withdrawal.

3. Any Party that withdraws from the Convention shall be considered as also having withdrawn from this Protocol.

Article 27

The original of this Protocol, of which the Arabic, Chinese, English, French, Russian and Spanish texts are equally authentic, shall be deposited with the Secretary-General of the United Nations.

Done at Kyoto this tenth day of December one thousand nine hundred and ninety-seven.

In witness whereof the undersigned, being duly authorized to that effect, have affixed their signatures to this Protocol on the dates indicated.

Annex A

Greenhouse gases

Carbon dioxide (CO2)
Methane (CH4)
Nitrous oxide (N2O)
Hydrofluorocarbons (HFCs)
Perfluorocarbons (PFCs)
Sulphur hexafluoride (SF6)

Sectors/source categories

- Energy
  - Fuel combustion
    - Energy industries
    - Manufacturing industries and construction
    - Transport
    - Other sectors
    - Other
  - Fugitive emissions from fuels
    - Solid fuels
    - Oil and natural gas
    - Other
- Industrial processes
  - Mineral products
  - Chemical industry
  - Metal production
  - Other production
  - Production of halocarbons and sulphur hexafluoride
  - Consumption of halocarbons and sulphur hexafluoride
  - Other
- Solvent and other product use
- Agriculture
  - Enteric fermentation
  - Manure management

Rice cultivation
Agricultural soils
Prescribed burning of savannas
Field burning of agricultural residues
Other

Waste

Solid waste disposal on land
Wastewater handling
Waste incineration
Other

Annex B

Party	Quantified emission limitation or reduction commitment (percentage of base year or period)
Australia	108
Austria	92
Belgium	92
Bulgaria*	92
Canada	94
Croatia*	95
Czech Republic*	92
Denmark	92
Estonia*	92
European Community	92
Finland	92
France	92
Germany	92
Greece	92
Hungary*	94
Iceland	110
Ireland	92
Italy	92
Japan	94
Latvia*	92
Liechtenstein	92
Lithuania*	92
Luxembourg	92
Monaco	92
Netherlands	92
New Zealand	100
Norway	101
Poland*	94
Portugal	92
Romania*	92

Russian Federation*	100
Slovakia*	92
Slovenia*	92
Spain	92
Sweden	92
Switzerland	92
Ukraine*	100
United Kingdom of Great Britain and Northern Ireland	92
United States of America	93

* Countries that are undergoing the process of transition to a market economy.

Total carbon dioxide emissions of Annex I Parties in 1990, for the purposes of Article 25 of the Kyoto Protocol[334]

Party	Emissions (Gg)	Percent
Australia	288,965	2.1
Austria	59,200	0.4
Belgium	113,405	0.8
Bulgaria	82,990	0.6
Canada	457,441	3.3
Czech Republic	169,514	1.2
Denmark	52,100	0.4
Estonia	37,797	0.3
Finland	53,900	0.4
France	366,536	2.7
Germany	1,012,443	7.4
Greece	82,100	0.6
Hungary	71,673	0.5
Iceland	2,172	0.0
Ireland	30,719	0.2
Italy	428,941	3.1
Japan	1,173,360	8.5
Latvia	22,976	0.2
Liechtenstein	208	0.0
Luxembourg	11,343	0.1
Monaco	71	0.0
Netherlands	167,600	1.2
New Zealand	25,530	0.2
Norway	35,533	0.3
Poland	414,930	3.0

[334] Data based on the information from the 34 Annex I Parties that submitted their first national communications on or before 11 December 1997, as compiled by the secretariat in several documents (A/AC.237/81; FCCC/CP/1996/12/Add.2 and FCCC/SB/1997/6). Some of the communications included data on CO2 emissions by sources and removals by sinks from land-use change and forestry, but since different ways of reporting were used these data are not included.

Portugal	42,148	0.3
Romania	171,103	1.2
Russian Federation	2,388,720	17.4
Slovakia	58,278	0.4
Spain	260,654	1.9
Sweden	61,256	0.4
Switzerland	43,600	0.3
United Kingdom of Great Britain and Northern Ireland	584,078	4.3
United States of America	4,957,022	36.1
Total	**13,728,306**	**100.0**

Source: FCCC/CP/1997/7/Add.1

vielfach dominieren.
Bd. 1, 1996, 160 S., 38,80 DM, br., ISBN 3-8258-2861-1

Manju Gupta; Klaus Voll
Indien: Geschichte, Politik und Gesellschaft der größten Demokratie der Welt
Bd. 2, Herbst 2001, 200 S., 38,80 DM, br., ISBN 3-8258-2865-4

Klaus Voll (Hrsg.)
Indische Außenpolitik
Bd. 3, Herbst 2001, 200 S., 38,80 DM, br., ISBN 3-8258-2866-2

Daniel Haas
Mit Sozialklauseln gegen Kinderarbeit? Das Beispiel der indischen Teppichproduktion
Das Thema "Sozialklauseln in internationalen Handelsverträgen" – seit dem Abschluß der *Uruguay-Runde* des *Allgemeinen Zoll- und Handelsabkommens* (GATT) wieder verstärkt in der Diskussion – steht im Spannungsfeld von Handels- und Entwicklungspolitik und muß im Kontext ungehemmter ökonomischer Globalisierung betrachtet werden. Können Sozialklauseln ein Regulativ für den sozial (und ökologisch) außer Rand und Band geratenen Weltmarkt sein? Können Sozialklauseln das entwicklungspolitische Instrumentarium zur Durchsetzung sozialer Mindeststandards sinnvoll ergänzen? Die vorliegende Arbeit betrachtet die Sozialklausel-Idee aus der Sicht eines Entwicklungslandes und richtet den Blick besonders auf die für die massive Ausbeutung von Kindern berüchtigte indische Teppichexportproduktion. Es werden konkrete Umsetzungsprobleme einer – wie auch immer gearteten – Verknüpfung des Zugangs zu internationalen Märkten mit der Einhaltung sozialer Mindeststandards in Exportbetrieben beleuchtet und Chancen sowie Risiken einer solchen Verknüpfung am vorliegenden Fall kritisch analysiert. Zu guter Letzt faßt der Autor die Reaktionen und Argumentationslinien von indischen Basisbewegungen, Gewerkschaftern und Regierungsvertretern zusammen und greift dabei auf reichhaltiges empirisches Material sowie eigene Interviews zurück.
Bd. 4, 1998, 152 S., 39,80 DM, br., ISBN 3-8258-3583-9

Sebastian Bersick
ASEM: Eine neue Qualität der Kooperation zwischen Europa und Asien
Das Asia-Europa Meeting (ASEM) ist in den zwei Jahren seines Bestehens zu einem zentralen Mechanismus der Zusammenarbeit zwischen Europa und Asien geworden. Parallel zu der bisherigen interregionalen Kooperation im Rahmen des EU-ASEAN-Dialogs sind innerhalb des ASEM-Prozesses, neben den Staaten der ASEAN, ebenso Japan, die Republik Korea und die Volksrepublik China Teilnehmer einer weiteren interregionalen Kooperationsinitiative. Gegenstand dieser Arbeit sind Genese und Entfaltung des ASEM-Prozesses sowie die Interessen, die von den beteiligten Akteuren verfolgt werden. Im Mittelpunkt der Analyse steht die Frage, ob und inwieweit das ASEM eine neue Form der Kooperation hervorgebracht hat und daher der eurpoäisch-asiatischen Zusammenarbeit eine neue Qualität verleiht. Mit dieser Untersuchung liegt die erste deutschsprachige Monographie über den ASEM-Prozeß vor, der jüngsten und zugleich ambitioniertesten Initiative innerhalb der europäisch-asiatischen Beziehungen.
Bd. 5, 1999, 128 S., 29,80 DM, br., ISBN 3-8258-4173-1

Heidelberger Studien zur Internationalen Politik

herausgegeben von Frank R. Pfetsch

Frank R. Pfetsch (Hrsg.)
Globales Konfliktpanorama 1990 – 1995
Mit Beiträgen von Peter Billing, Andreas Busch, Christof Hartmann, Sabine Klotz, Marianne Rinza, Christoph Rohloff, Hardi Schindler und Bernhard J. Trautner
Bd. 1, 1996, 317 S., 38,80 DM, br., ISBN 3-8258-3000-4

Bernhard J. Trautner
Konstruktive Konfliktbearbeitung im Vorderen und Mittleren Orient
Ansätze der Deeskalation und Beilegung nationaler und internationaler Konflikte 1945 – 1995
Bd. 2, 1997, 416 S., 59,80 DM, br., ISBN 3-89473-962-2

Hardi Schindler
Konflikte in Südamerika
In der Friedens- und Konfliktforschung ist die Region Südamerika vernachlässigt und unterrepräsentiert. Dabei vollzog sich dort ein bemerkenswerter Wandel weg von politischer Instabilität mit Putschen, Militärdiktaturen, Guerillakriegen und ökonomischem Niedergang hin zur Demokratie, wirtschaftlicher Entwicklung und regionaler Integration. Die abgegrenzte geographische Lage Südamerikas, die relative religiös-kulturelle Homogenität und die gemeinsame Geschichte begünstigen eine konflikttheoretische Konzentration auf besondere Schwerpunkte.
Die Arbeit beschreibt und analysiert die 27 politischen Konflikte, die in Südamerika seit dem Ende des Zweiten Weltkrieges zu verzeichnen